THE
EVOLUTION HOAX
EXPOSED

THE
EVOLUTION HOAX
EXPOSED

(Former Title: *Why Colleges Breed Communists*)

by

A. N. FIELD

TAN BOOKS AND PUBLISHERS, INC.
Rockford, Illinois 61105

First published in 1941 under the title *Why Colleges Breed Communists*. This edition reprinted from the 1971 edition of Christian Book Club of America, Hawthorne, California.

First TAN printing, December 1971	10,000
Second TAN printing, May 1976	9,000
Third TAN printing, January 1984	5,000

ISBN: 0-89555-049-0

Printed and bound in the United States of America.

TAN BOOKS AND PUBLISHERS, INC.
P.O. Box 424
Rockford, Illinois 61105

1971

CONTENTS

6

CONTENTS

PREFACE

Evolutionism is the subject which the author examines. It is a subject the implications of which are, for some, obscured by the rodomontade of certain scientists. More dangerously, the true character of evolutionism is clouded for many by the vapourings of professional religionists, and in particular by such as choose to dispense that peculiar mixture of clergianity and Marxism known as the Social Gospel. In the mass media of England, the T.V., the press and the wireless, evolutionism is sacrosanct, and its speculative basis must not be discussed. The more vocal propagandists of the cult, however, have been granted access to the schools and universities. It is as paid educationalists that these are now licensed to press their atheism upon captive audiences, and, where and whenever possible, to overthrow the faith of children.

In this compact survey, Mr. Field shows evolutionism to be a scientific fraud. He brings forward the facts which enabled him to recognise the modern university college as a machine for de-Christianising and demoralising the community.

In his chapter, *Evolution's Offspring*, the author opens up a very fruitful line of research, and reveals the origin of much of the spiritual and intellectual unrest of our times. This is generated by the wilful abandonment of Scripture; yet there are signs that the wheel is coming full circle. Again, one here, one there, is prompted to return to, and to dare to believe what God has written.

May this book play some part in encouraging that return.

THE PUBLISHERS.

Chapter I

THEORY VERSUS REALITY

THE PURPOSE of this book is to show the falsity of a theory which for three-quarters of a century has dominated all British thought. Nominally, the theory of organic evolution is no more than a proposition in biological science. Actually, it is the parent and support of a whole host of subsidiary theories—political, economic, and scientific—all accepting as established truth things which are wholly unproven, and many of which are demonstrably untrue.

On evolution's twenty-first birthday in 1880, Thomas Huxley, chief propagandist of the Darwinian gospel, remarked that it was "the customary fate of new truths to begin as heresies, and to end as superstitions (Huxley's *Life and Letters*, ii, 12). Evolution, disguised as truth, has run this full course. In the light of all the facts it is no overstatement to say that it has made the universities and higher educational institutions of the British Empire, not centres of enlightenment but haunts of superstition and intellectual darkness. The theory of evolution is only kept going by constant distortion of observed fact, and its main result is wrong-headed thinking on all aspects of life.

In the universities of Britain, the British Empire, and the United States a strong under-current of subversive influence has been operating on the minds of students for many years past. Many people who have never had occasion to look into the matter closely regard this as due to nothing more than misguided ebullitions of youthful enthusiasm. There is evidence, however, of organised effort over many years in the work of deliberately pumping subversion into Britain's universities. Even a generation back it was recognised in New Zealand that the little trickle of university professors from Britain was bringing with it a steady insinuation of materialistic and subversive ideas into the minds of students. Since the outbreak of the present war public feeling on the matter has found emphatic expression in some parts of New Zealand. Evolutionist teaching prepares the ground for subversion.

It is well recognised that this mental infection is apt to colour the whole outlook on life of those who suffer it. Two professors of the University of London in 1933 were so impressed by this that they set out to collect statistics as to how far students supporting one "advanced," "progressive," or "unorthodox" movement tended to support other such movements. The questionnaire did not inquire if the students believed in evolution, for this is nowadays taken for granted. According to the London *Patriot* of May 25, 1933, the numerous movements listed in the professorial questionnaire included birth control, abortion, sexual freedom, new education (without coercion), rationalism (atheism), nudism, psycho-analysis, anarchism, communism, socialism, refusal of military service, sterilisation, etc., etc. The universities were not established as hotbeds of propaganda for movements of the above character, but have become such.

In the United States in 1934 a somewhat similar questionnaire was circulated by an undergraduate organisation of Harvard University, the numerous questions in the forms distributed being, however, almost wholly on psycho-analytic sex lines with inquiry into students' views and habits (*Boston Evening Transcript*, March 14 and 15, 1934). These two inquiries bear sufficient testimony to a similar trend of infection in leading educational institutions in both Britain and the United States.

The practical outcome of university education today is the production of people with a boundless belief in all manner of unverified, and often unverifiable, theories. At the same time the general public is inspired with an equally boundless, and equally groundless, belief in anything labelled as Science. Experience is thrust out of the window as a useless teacher, and some little tom-noddy of a university graduate with a bagful of theories is blindly entrusted with the task of remaking heaven and earth.

In world affairs we see the fruits of modern university education in the present difficulties in which the British Empire finds itself involved, and from which the fortitude and resolution of the common people are left to extricate it. Throughout the nineteenth century, before the theorists took charge, British foreign policy was based on intelligent and practical principles. Ample armed force and minimum interference in European affairs was the rule. And long stretches of almost unbroken peace were the result.

At the end of the World War in 1918 our educated theorists got the bit properly between their teeth, and the edifice known as the League of Nations was exactly the sort of product to be expected from university-minded people. This scheme for the manufacture of peace on earth was all theory without any working parts whatever. Its principal promoter in the British Ministry during the last war got a Foreign Office report in 1916 on the draft plan, which report by Sir Eyre Crowe duly pointed out in detail that the projected League would do everything except operate as desired. This trifling shortcoming, however, was held in no way to detract from the theoretical beauty of the plan for remaking mankind. This text of this instructive report will be found in the Lloyd George *War Memoirs* (vol. iii).

In their worship of this Palace of Talk at Geneva our evolutionist university-minded intellectuals were prepared to neglect and sacrifice every British interest. The clamour of these theorists resulted in Britain first throwing away her arms, and then entangling herself in every possible and impossible direction in other people's business in Europe and elsewhere.

After four years of suffering and endurance by the common people, the British had emerged victorious from the last war. All the fruits of victory were flung away by the theorists. An amazing financial policy was pursued at the bidding of a private corporation of secret and possibly foreign ownership : and the result was unemployed workers by the million over a period of twenty years—exactly as was predicted in the London *Times* in 1918 on the policy being first mooted. No effort worth speaking of was made to develop the nation's world-wide heritage. A great part of the time of the political heads was taken up in rushing from one international conference to the next, and signing pact after pact, each of which duly proved worthless almost before the ink was dry on it.

In the end, this twenty years of unreal politics based on unreal education collapsed like the house of cards it was. Britain found herself plunged into war under more disadvantageous circumstances in point of equipment and allies then ever before in her history. In this struggle the mass of the nation as before is exhibiting the high qualities of the British race.

As for the intellectuals, they have made the war the occasion for producing an even more flamboyant theoretical construction than their League of Nations Plan of the last war. Under the name of Federal Union this proposes a restoration of the gold standard for the benefit of the international financiers owning the world's gold stock; the dissolution of the British Empire; and, for all practical purposes, its virtual absorption by the United States. This remarkable project the present writer hopes to review at a later date.

Such are the fruits of our theory-mad age. And the fountain-head of these dreams and imaginings, divorced from reality is undoubtedly the theory of organic evolution produced by Charles Darwin just over eighty years ago. This is the grand river of falsity and corruption from which all sides of national life have been irrigated with the waters of untruth. The effects of this theory are so far-reaching that they deserve the attention of all. In the following pages we will trace out the present position of the case for evolution, the origin and development of the theory, and some of its consequences.

Chapter II

THE SKELETON IN THE CUPBOARD

WITH oaks to be seen sprouting from acorns, grubs turning into butterflies, and chickens pecking their way out of eggs, it is not surprising that human fancy from an early date toyed with the notion of one kind of living thing being transformed into some other kind. This idea has been the stock-in-trade of folk-lore and fairy tales in all ages and all lands. It was the achievement of Charles Darwin to make it the foundation of modern biological science.

At the end of the eighteenth century there occurred that great event known as the French Revolution, described in various quarters as a landmark in the liberation of the human spirit. Incidentally, the student may learn from Alison's *History of Europe* how in the course of this episode the mob in the streets of Paris roasted and ate the bodies of the massacred Swiss Guard of the royal palace on August 10, 1792, and how fifteen months later the multitude assembled in the Cathedral of Notre Dame to worship the Goddess of Reason, personified by an actress, also well known to the public in another capacity, placed naked by Government decree upon the altar of the French Westminster Abbey. State and people having alike discarded Christianity as outworn superstition, attention was directed in scientific circles to discovering how the world had come into being without intervention of the Almighty.

Modern evolutionist theory dates from the writings of two French scientists of this period, the Comte de Buffon and M. Lamarck, the latter of whom propounded a theory of evolution by inheritance or acquired characters. Evolutionist ideas were also advanced in England by a medical man and author of Derby, Dr. Erasmus Darwin, correspondent and admirer of Jean Jacques Rousseau, chief philosopher of the French revolutionary era. Ideas of evolution began to float about, and in 1813, Dr. W. C. Wells aired the subject before the Royal Society in London.

Scientists in the first half of the nineteenth century were by no means all believers in Evolution. The Creationists were in a great majority, and the Evolutionists lacked a plausible theory of how evolution had occurred. This was all altered on the appearance of Charles Darwin's *Origin of Species* in 1859.

Darwin's theory was that all species had evolved from primal protoplasm by Natural Selection, or survival of the fittest in the struggle for existence. As lambs still continued to appear with tails, despite the fact that their parents had had theirs cut off for generations back, Lamarck's idea of evolution by inheritance of acquired characters had failed to convince. Darwin not only provided a more attractive theory but he elaborated it in a big book. He directed attention to the way breeders of domestic animals and plants got new varieties by selective breeding. He quoted instances of small variations occurring in all directions in living things. He argued that the rate of multiplication of living things produced an intense struggle for existence exterminating those with unfavourable variations and permitting those with favourable variations to flourish and increase.

The idea of evolution as thus propounded met with immediate and astounding success. Not all scientists accepted it by any means, but it caught on and disbelievers were soon shouted down. In his presidential address to the British Association for the Advancement of Science in 1937 Sir Edward Poulton said the records showed that Dr. Wright back in 1881 was the last person to express disbelief in evolution at this British scientists' parliament. The article on evolution in the current (1929) edition of the *Encyclopaedia Britannica* affirms evolution to be an established fact supported by "overwhelming" evidence. In current newspaper and magazine literature evolution is similarly treated as a fact beyond dispute, and in colleges and schools is usually rammed down the pupils' throats as incontrovertible truth.

Nevertheless, there is a small, slight hiatus in the argument. There is, unfortunately for evolutionists, not a shred of evidence of any living thing ever evolving into some different kind of living thing capable of breeding but infertile with its parent stock. All that breeding experiments have produced is mere varieties fertile with their parent stock, or else sterile hybrids, incapable of breeding, such as the mule produced by a cross between horse and donkey. All living things go on obstinately producing young after their own kind and no other kind. Evolution has to show that living things can break through their natural breeding limits. And this is just what evolution has been quite unable to show.

This small defect in an otherwise pleasing theory Darwin glossed over in his books. Nevertheless, Darwin admitted in the introduction to his *Origin of Species* that evolution as a scientific theory "would be unsatisfactory, until it could be shown how the innumerable species inhabiting this world have been modified." Before he got to the end of his 700 pages Darwin ignored this requirement, for, without meeting it, he declared himself in his concluding chapter "thoroughly convinced" of evolution.

Huxley, who from the outset constituted himself the chief propagandist of Darwinism, was more logical. He made no bones about the total absence of any actual proof of evolution. In fact, he greatly annoyed Darwin by harping on the point, as anyone who cares to peruse the five volumes of Darwin's letters and the two volumes of Huxley's may discover for himself. " My God," wrote Darwin to Huxley in 1862, "is not the case difficult enough without its being, as I must think, falsely made more difficult. I believe it is all my own fault—my own d——d candour . . ." (*More Letters of Charles Darwin*, i, 230).

This outburst was in consequence of Huxley having pointed out in his lectures and books that so far it had not been possible by selective breeding to produce a form capable of breeding but infertile with the parent stock. Huxley, in admitting this lack of evidence, said in his Edinburgh lectures in 1861 that if it could be shown that such failure was "the necessary and inevitable result of all experiments" he would hold Mr. Darwin's hypothesis to be "utterly shattered." (*Man's Place in Nature*, Everyman edition, p. 256). He added, however, that he looked for early proof to be forthcoming. In a letter to Darwin, Huxley said he told his students that he was satisfied that twenty years' scientific breeding experiments with pigeons would provide the necessary proof (Huxley's *Life and Letters*, i, 195-6).

Although so heatedly rebuking Huxley in 1862, Darwin himself eight months later, privately admitted in a letter to Dr. Bentham the total absence of any proof of evolution. This is what he wrote under date of May 22, 1863 : "In fact belief in Natural Selection must at present be grounded entirely on general considerations. . . . When we descend to details, we can prove that no one species has changed (*i.e.*, we cannot prove that a single species has changed); nor can we prove that the supposed changes are beneficial, which is the groundwork of the theory. Nor can we explain why some species have changed and others have not" (Darwin's *Life and Letters*, iii, 25).

Darwin died in 1882. Huxley died in 1895. Four years before he died Huxley wrote to Professor Romanes that evolution still stood without the evidence on which he had insisted thirty years before. He had always insisted, he said, on "the logical incompleteness of the theory so long as it was not backed by experimental proof" (Huxley's *Life and Letters*, ii, 291).

We now pass onwards another thirty years; and we find the noted English evolutionist, Professor Sir William Bateson, acknowledging exactly the same total absence of any proof of evolution. This is what he said to the congress of the American Association for the Advancement of Science in Toronto in 1921 :—

"When students of other sciences ask us what is now currently believed about the origin of species we have no clear answer to give. . .

The conclusion in which we were brought up, that species are a product of summation of variations, ignored the chief attribute of species, that the product of their crosses is frequently sterile in great or less degree. Huxley very early in the debate pointed out this grave defect in the evidence, but before breeding researches had been made on a large scale no one felt the objection to be serious. Extended work might be trusted to supply the deficiency. It has not done so, and the significance of the negative evidence can no longer be denied" (*Nature*, Ap. 29, 1922).

Six years later another prominent evolutionist, Professor J. B. S. Haldane, in his book *Possible Worlds* (p. 38) said in 1927 : "The barrier of inter-specific sterility is the most serious argument against Darwin's Organic Evolution." It is equally a barrier against any other kind of organic evolution.

In 1931 we find the great Professor H. F. Osborn, of the United States, described by Britain's Royal Society as the greatest palaeontologist of the day, making the following remarkable statement to a congress of the British Association :—

"We are more at a loss than ever to understand the causes of evolution. One after another the Buffonian, Lamarckian, Darwinian, Weissmannian, and De Vriesian theories of causation have collapsed ... All that we can say at present is that Nature does not waste time or effort with chance or fortuity or experiment, but that she proceeds directly and creatively to her marvellous adaptive ends of biomechanism" (*Nature*, September 28, 1931).

This is an admission by one of the high priests of science that all theories of evolution have collapsed. Buffon propounded a general theory of evolution; Lamarck, tutor to Buffon's children, followed with an idea of evolution by inheritance of acquired characters; Darwin advanced the idea of gradual small changes by natural selection; Weissman put all the emphasis on the germ plasm; and De Vries put forward the idea of evolution by mutations or sudden large variations.

Two years later on again Professor James Ritchie, the great zoologist of Edinburgh University, wrote in *Nature* of September 30, 1933 : "The problem of the origin of species seems to be as far from solution as ever." In September, 1939, Professor Ritchie delivered the presidential address to the zoological section of the British Association and had nothing further to report. "The existence of life," he said, "must be considered as an elementary fact which cannot be explained," and, admitting life, the biologist "may build up a whole body of biological theory . . . logical in the logic of probability . . ." (*Nature*, September 23, 1939). Everything was down to mere shadowy "probability."

The above series of pronouncements by front-rank biologists covers the entire period of eighty years since the first proclaiming of the Darwinian gospel. There is no more vestige of proof of evolution today than there was in those early days when Darwin privately, and Huxley openly, admitted its total absence.

Claims are made from time to time of the production by experiment of new species of living things, but they rapidly drop out of sight and the above series of statements at scientific headquarters is sufficient evidence that no such claim has survived examination. This evidence is vital to the evolution theory, and if it were forthcoming we may be quite sure it would be proclaimed from the house-tops for all the world to hear.

If this evidence is lacking it is not for want of seeking it. For example, a whole literature, so extensive that a bibliography of it was recently published, has grown up about the breeding experiments with the pumice fly *Drosophila melanogaster*. Mr. Douglas Dewar, a Fellow of the Zoological Society and one of the few British biologists rejecting evolution, in his *Challenge to Evolutionists* (pp. 20-21) relates how in 1910 Morgan and his collaborators hit upon the idea of experimenting with this quick-breeding fly.

This obliging little creature produces 25 generations a year at ordinary temperatures and more at higher temperatures. Over 800 generations of it have been bred with the object of transforming into something that is not a *Drosophila melanogaster*. It would take 20,000 years to get as many generations of human beings. Every device has been applied to this fly to make it vary its breeding. In 1927 it was discovered that by exposing it to X-rays the rate at which mutations, or marked variations, occurred could be increased by 15,000 per cent.

These breeding experiments are stated to have resulted in the production of some 400 varieties of this fly, some of them monstrosities, and some differing more from the parent form than the other wild species of *Drosophila* differ from one another. Nevertheless, all these varieties (unless they are too imperfect to breed at all) are stated to breed freely with the parent stock, whereas the different wild species of *Drosophila* on the rare occasions when they can be induced to cross, either yield no offspring at all or sterile hybrids. Immutability of species, like a mysterious angel with flaming sword, stands barring the way to the evolutionist Garden of Eden.

Summed up, the position is that there is no evidence of any interbreeding community of living things being able to change its breeding and become transformed into some different kind of thing infertile with the original stock. Evolution asserts that all species came into being in this way. And evolution is wholly unable to provide any vestige of proof of its assertion. Belief in evolution today must thus rest on "general considerations," just as Darwin privately confessed was the case away back in 1863. In our next three chapters we shall discover how these general considerations—the cloud capp'd towers and gorgeous palaces of evolution—melt away like the baseless fabric of a vision when peered at too curiously.

Chapter III

EVOLUTION'S CASE TODAY

THE first thing noticeable about modern presentations of the case for evolution is the widely different ideas held by various evolutionists as to what constitutes proof of a proposition. If we make our starting point the article on evolution in the current (1929) edition of Britain's standard reference book the *Encyclopaedia Britannica*, we read that evolution is supported by "overwhelming" evidence. If we turn to certain of the recent presidential addresses to the biological sections of the British Association, we shall find scientists there affirming on exactly the same evidence that evolution is not a proved fact but purely a matter of faith.

We live in an age of propaganda circulated throughout the world from obscure sources for obscure ends. Perusal of some articles in the successive post-war editions of the *Encyclopaedia Britannica* shows in various cases a haphazard reversal of view between one edition and the next, and in particular a change from a factual to a propagandist view of evolution in the three brief years between the thirteenth edition of this work of reference in 1926 and the fourteenth in 1929.

In the 1929 edition two eminent biologists combine to write the general article on evolution. There is a brief, very positive and very dogmatic introduction by the biological editor of the edition, Professor Julian Huxley, then professor of physiology at the Royal Institution; but when it comes down to cold hard facts a great part of the positiveness fades away in the main part of the article written by Professor E. S. Goodrich, professor of zoology and comparative anatomy at Oxford University.

Both these ardent evolutionists make weak starts. The first item is an announcement by Professor Huxley that "among competent biologists and geologists there is not a single one who is not convinced that evolution has occurred and is occurring," and Professor Goodrich on taking up the running immediately asserts that "it is now universally held by all competent biologists" that evolution is a fact.

These statements are weak for two reasons. In the first place they suffer from the defect of being untrue. There are some fully competent biologists and geologists who have publicly rejected the entire theory of evolution as baseless. A list of them will be found on pp. 102-105. In the second place, the statements are weak because the truth of a scientific proposition is not to be established by counting noses. Professor Julian Huxley's own eminent grandfather expressed himself very freely on this matter of nose-counting. "Government by average opinion," he wrote, "is merely a circuitous method of going to the devil. Those who profess to lead but in fact slavishly follow this average opinion, are simply the fastest runners and loudest speakers in the herd which is rushing blindly to its destruction" (Huxley's *Life*, ii, 125).

When theologians state that most people believe in the existence of God Almighty, evolutionsts like Professor Julian Huxley tell us this is merely evidence of the prevalence of superstition and credulity. On the other hand, credulity in scientific circles apparently turns fancy into fact. Besides being unreliable in their nose-counting, these two eminent encyclopaedists are illogical in their deductions from it.

However, it appears that there is secondary support for evolution beyond the mass of scientific noses upholding the theory. Professor Huxley assures us in the encylopaedia that "by now the evidence is overwhelming." The evidence is as follows :—

(1) Fossils, which are stated to provide "complete proof,"

(2) Vestigial organs, that is to say, useless parts remaining from ancestral forms of being;

(3) Embryology, showing the embryo developing through ancestral forms of the species.

In addition, the general plan of plants and animals is said to bear witness to a common descent. Their geographical distribution is described as easy to explain on evolutionist lines, but difficult otherwise. Also, evolution is held to have "pragmatic value" in explaining things.

The foregoing are simply the "general considerations" which Darwin, as we have seen, referred to in 1863 in privately admitting the absence of any rag of proof that any living thing could change into any other kind of living thing infertile with the parent stock. Professor Julian Huxley passes over this total absence of real proof of evolution without notice.

The decline of the *Encyclopaedia Britannica* from a factual to a propagandist work of reference is evidenced by referring to the article on embryology appearing in the thirteenth edition three short years before Professor Julian Huxley was holding forth in the 1929 edition. This thirteenth edition article was written by Professor Adam Sedgwick, professor of zoology at the Imperial College of Science and Technology. Professor Sedgwick dismisses as baseless the idea that fossils, embryology, and vestigial organs provide any "proof" at all of evolution. In discussing the Darwinian theory as elaborated by

Haeckel that the embryo in developing recapitulates the ancestral history of the species, Professor Sedgwick says of this class of "evidence" :

"When we come to look for the facts upon which it is based, we find that they are non-existent, for the ancestors of all living animals are dead, and we have no means of knowing what they were like. It is true there are fossil remains of animals which have lived, but these are so imperfect as to be practically useless for present requirements. Moreover, if they were perfectly preserved, there would be no evidence to show that they are the ancestors of animals now living. They might have been animals which have become extinct and left no descendants.

"Thus the explanation ordinarily given of the embryonic structures referred to is purely a deduction from the evolution theory. Indeed, it is even less than this, for all that can be said is something of this kind : if the evolution theory is true, then it is conceivable. (Note : not 'it is certain,' or even 'it is probable') that the reason why the embryo of a bird passes through a stage in which its pharynx bears some resemblance to that of a fish is that a remote ancestor of the bird possessed a pharynx with lateral apertures, such as are at present found in fishes." Professor Sedgwick remarks incidentally that although fishes have teeth, no teeth are to be found in bird embryos.

In Professor Sedgwick's view what Professor Julian Huxley calls "overwhelming" evidence is no evidence at all. The evolutionist can find a succession of fossil remains of different types of animals in different geological strata : but that present animals are the descendants of extinct ones is pure assumption. The whole evolutionist case is made up of one assumption placed on top of another assumption. All is guesswork from start to finish.

Before going on to the main evolutionist line of "complete proof" in the fossils, the minor counts are worth looking over. With respect to the common plan of plants and animals, this amounts to little. All plants and animals have to live in a common environment of land, air, and water, and a common plan is just as much to be expected on Creationist as on Evolutionist belief. As to geographical distribution, Darwin wrote to Hooker that to get his species about the world, he was always ready to raise up "former immense tracts of land in oceans if any case required it in eminent degree," adding : " . . . at present I much prefer land in Antarctic regions . . . you have thus to invent much less land, and that more central . . ." (*More Letters*, i, 115). If Darwin could do this, the Almighty should be able to manage it also. As to evolution's pragmatic value in explaining things, one finds an evolutionist author, Mr. A. Beebe, quoted as writing quite seriously in his book *The Bird* : "The idea of miraculous change which is supposed to be an exclusive prerogative of fairy tales is a common phenomenon of evolution." Does "pragmatic value" simply mean that in scientific circles an untrue explanation is considered better than no explanation?

Professor Goodrich, in coming down to the details in the fourteenth edition of the encyclopaedia, speedily loses the easy positive touch of Professor Huxley in opening the evolution article. Professor Goodrich begins by dispensing with the services of the Almighty altogether in his evolutionist Garden of Eden. Everything came from protoplasm, and "there must have been a time" when protoplasm first appeared. It "must be supposed" that inorganic substances started forming compounds, and that some of these kept on reforming themselves, and "once they started on this trick" they "would inevitably tend" to perpetuate themselves. These things "probably occurred" in the sea.

The actual fact of the matter is that neither Professor E. S. Goodrich, F.R.S., nor any other scientist, can produce for examination any inorganic substances which keep on forming compounds indefinitely and turning into living organisms. All the many attempts of scientists to produce living matter from non-living matter has been a dead flat failure. What Professor Goodrich talks of is no more producible than are Hans Andersen's witch with the tinder-box and dog with eyes as big as mill-wheels. The main difference is that Hans Andersen's flights of fancy are easier reading than professorial jargon.

Vestigial organs in animals and plants are stated by Professor Goodrich to be "numberless." The human vermiform appendix and the splint bones in horses' legs are two much-quoted vestigial organs. However, the professor proceeds in the encylopaedia with the following remarkable statements about such organs : "Unless they have been adapted to fulfil some new function, they are apt to diminish and disappear . . . It is doubtful whether any useless parts are ever preserved for long unless they are insignificant, and many of the so-called vestigial organs are now known to fulfil important functions."

According to Darwin, evolution is proved (or nearly so) by the presence of organs "bearing the plain stamp of inutility" and "imperfect and useless." On the theory of creation, he argued, the presence of useless organs was inexplicable. On the theory of evolution, they were explainable as atrophied survivals from past forms of being. When Professor Goodrich talks of many vestigial organs fulfilling useful functions he at once knocks completely to pieces any claims such organs have as evidence of evolution. Grant them any use and they wholly cease to be vestigial organs in the Darwinian sense of useless survivals. The modern evolutionist tries to eat his cake and have it.

Mr. Douglas Dewar, F.Z.S., an anti-evolutionist, discusses these vestigial organs at length in his *Difficulties of the Evolution Theory* and *More Difficulties of the Evolution Theory*. He points out that the number of supposedly useless organs decreases as biological know-ledge increases, and he suggests that those which remain may represent no more than the measure of our ignorance. If biologists had not been so busy hunting for useless organs to prove evolution, the use of many might have been discovered already.

Evolutionists describe the splint bones in horses as vestigial remains of extra toes. Mr. Dewar in his *More Difficulties* (p. 54) quotes Hayes, an authority on the horse, as stating that these bones (1) strengthen the leg, (2) serve as an attachment for certain muscles, and (3) in conjunction with the canon bone form a groove in which lies the upper part of the suspensory ligament supporting the fetlock and counteracting the effects of weight. How much is left of their vestigial uselessness?

Since King Edward VII had his coronation postponed for a year in order to have his vermiform appendix removed, vast numbers of lesser persons have discovered that they can part with this portion of their anatomy without immediately fatal results. Mr. Dewar, in his *Man, a Special Creation*, notes Dr. Le Gros Clark, professor of anatomy at St. Thomas's Hospital, London, saying in 1934 in his book *Early Forerunners of Man* (p. 205): "The significance of the vermiform appendix is still quite obscure, but in view of its rich blood supply it is almost certainly correct to regard it as a specialised and not a degenerate organ."

Mr. Dewar remarks that while Darwin wrote in a general way about nascent organs, no evolutionist has been able to point in either extinct or existing forms of life to any nascent organ in course of development; yet if evolution is a fact all organs must have had rudimentary beginnings. The complete absence of nascent organs is usually passed over in silence in evolutionist literature. Enough hurdles have to be jumped without looking for more.

Embryology is affirmed by Professor Goodrich in the encyclopaedia to afford "strong evidence" of evolution; but he admits that Haeckel's law of recapitulation is "a gross exaggeration." In the 1929 encyclopaedia article on embryology by Professor D. M. S. Watson, professor of zoology at London University, we learn that research to confirm Haeckel "can scarcely be said to have succeeded in its original aim."

In 1938 there was published at Oxford a book of essays entitled *Evolution*, and written by colleagues and former students of Professor Goodrich in commemoration of his seventieth birthday. The editor was Mr. G. R. De Beer, senior demonstrator in zoology and lecturer in embryology at Oxford. Here is all that Mr. De Beer will allow for embryology as "proof" of evolution:

". . . Very soon, in their enthusiasm for the great new revelation (evolution), biological students were making embryological facts subservient to their evolutionary theories. . . Thus arose the famous theory of recapitulation . . . as is so often the case with half-truths, this theory enjoyed wide acceptance . . . Nevertheless, it must be realised that the theory contained a fallacy which for two reasons impeded the progress of biological work and thought . . . In many cases it can be proved that the developmental history cannot represent the phylogenetic (species) history" (pp. 57-58).

This is open admission by an ardent evolutionist of one of the chief counts brought by anti-evolutionists. They complain, and on good grounds, that the theory of evolution has led to continuous and wholesale distortion of observed fact in order to make it fit in with evolutionist preconceptions. To such an extent has this proceeded that in modern scientific literature, especially the popular variety, it is impossible to distinguish between what has been observed and what is speculative embellishment. Fact and fairy-tale are muddled up indiscriminately.

According to Mr. De Beer, the most that can be said for embryological evidence of evolution is that "the structure of the adult ancestral form may sometimes be *inferred* (his italics) from that of the developmental stages of its descendants " (p.61). In other cases "little or no information" may be gleaned. On top of this, as Professor Sedgwick points out, nobody knows what the ancestral forms of any animal were, and what little remains of embryological "proof" of evolution thus subsides into moonshine.

Darwin wrote in the *Origin of Species* of gill-slits and a tail in the human embryo as evidence of fish and animal ancestry. Mr. Dewar and other evolutionist and anti-evolutionist writers point out that there are no slits at all, but simply depressions. These have nothing to do with breathing arrangements but develop into tissues connected with the ear, lower lip, tongue, cheek, and various other things. As for the embryonic tail, this bends inwards and forms the attachment for various muscles and also gives additional support to man's internal organs necessitated by his upright posture. Mr. Dewar remarks in his *More Difficulties* (p. 36) that in the human embryo a length of intestine also projects from the body for a period, but so far evolution has not claimed this as evidence of anything. All the indications, he states, are that the embryo wastes no time in meandering about repeating ancestral history, but develops in the most direct and expeditious manner possible in the circumstances.

Enough has been quoted in admissions from evolutionist sources to show that the miscellaneous exhibits in the shop window come down to very little indeed. None of the items provides any proof of anything, and the most that can be said of any is that it provides a basis for speculation. We now turn to the last remaining line of evidence, the fossils.

Chapter IV

FOOTPRINTS ON THE SANDS OF TIME

IT has been said that a man studying the fossilised remains of plants and animals can take away from them any theory he brings with him at the outset. In other words, the fossils themselves are one thing, and the inferences drawn from them are another matter altogether. In dealing with this part of evolution's evidence, it is thus advisable to bear in mind just what the rocks show. To begin with, the geologists divide the succession of strata in which fossils are found into three great ages :

(1)   The Palaeozoic Era (Ancient Life), the age of shells, corals, crabs, lobsters, and later the fishes; on land, scorpions and insects appear, and at the end of the period the earliest reptiles. Seaweeds are the earliest plants, followed by mosses and ferns on land, some growing to giant size in the swamp forests of the world's coal-bearing rocks of this era.

(2)   The Mesozoic Era (MiddleLife), the age notably of giant reptiles, in which the birds later appear, all this life reproducing by spawn or eggs; the first mammals, producing their young alive, also appear; seed-plants and trees are also found.

(3)   The Cainozoic Era (Newer Life), the age of mammals and of flowering plants.

These eras, also known as Primary, Secondary and Tertiary, are sub-divided into fifteen periods, the earliest period of the Palaeozoic era being known as the Cambrian. Each of the three geological ages is separated from the next by a marked physical break, with upheavals or outbreaks of volcanic activity.

The evolutionist contention is that the later forms of life seen in the rocks are the descendants of the earlier forms. The evolutionist infers this. The fossils themselves show no more than the order of succession in which the different types of plants and animals appeared. Transitional forms which must have existed if evolution is a fact, and for which names were allotted in advance after evolution became fashionable—pro-this and pro-that—have failed to appear, although the whole earth has been ransacked for them during the past eighty years.

The anti-evolutionists contend that the fossils do not in the least support the idea of evolution of species from a common stock. Even an evolutionist like Professor Max Westenhofer writes in 1937 in his *Research and Progress* (iii, 92) : "All the larger groups of animals, *e.g.*, fishes, amphibians, reptiles, mammals seem to have appeared suddenly on the earth, spreading themselves, so to speak, in an explosive manner in their various shapes and forms. Nowhere is one able to observe or prove the transition of one species into another, variation only being possible within the species themselves" (as quoted in Dewar's *More Difficulties*, p. 94).

Dr. W. Bell Dawson, F.R.C.S., a Laureate of the French Academy of Sciences, and a well-known Canadian geologist, says of the fossils : "This sequence is evidently the same as in Genesis; for in both, the creatures that swarm in the sea come first, and the land animals last. When each type of creature comes into being, it continues to the present day; as, for example, the sea shells, the crabs, and the reptiles; but in each type there is a magnificence in the past from which it has now deteriorated. Many ancient species are identical with forms still living; and many organs of their bodies, such as the claw of the lobster or the multiple eye of the crab, are precisely the same as in the earliest ages without any sign of improvement. Whole categories of facts such as these, seem to be lost sight of by those whose vision is obscured by evolutionary theories; for they run counter to any conception of evolution" (*The Bible Confirmed by Science*, pp. 73-74).

Another geologist, Joseph Le Conte, says : "The evidence of Geology today is that species seem to come into existence suddenly and in full perfection, remain substantially unchanged during the terms of their existence, and pass away in full perfection Other species take their place, apparently by substitution, not by transmutation" (as quoted by Dawson, *op. cit.*, p. 75).

Evolution's first hurdle in the fossils is the sudden bursting into view in the Cambrian rocks of a highly developed aquatic fauna and flora. Scientists hold that the time required for life to reach the stage there exhibited was greater than for all the subsequent development. Yet they are unable to find evidence of this first half of evolution. In the earlier Pre-Cambrian rocks there is nothing save a few vague markings claimed by their discoverers as fossils, but everyone of which is disputed.

Mr. Dewar, in *More Difficulties of the Evolution Theory* (p. 114) states that, except for Walcott's supposed Beltina in North America and David and Tillyard's supposed Eurypterid in Australia (both of them kinds of sea-scorpions), the rest of the few alleged pre-Cambrian fossils are either supposed secretions, or marks made by animals or plants, or organisms that fit into no known group of animals or plants.

The president of the American Palaeontological Society said in 1935 that Walcott's find could not be accepted until verified by further discoveries; and Britain's leading scientific journal *Nature* of December 12, 1936, described the Australian find as "unconvincing." Even if the whole little handful of these disputed Pre-Cambrian fossils is accepted they still throw no light on the evolution of the mass of life abruptly appearing in the Cambrian rocks.

Evolution supposes that fishes turned into reptiles, and reptiles into birds, elephants, and men and so on. The innumerable transitional forms demanded by the theory cannot be found. At one time much was heard of the fossils, *Archaeopteryx*, a fully feathered bird with teeth, and *Archaeornis*, as links between bird and reptile, but even the Oxford University *Evolution* book of 1938 dismisses these as "indubitably birds" (p. 322). In Professor J. B. Pettigrew's *Design in Nature* (p. 207), Huxley is quoted as rejecting *Archaeopteryx* as a missing link as far back as 1876.

Mr. B. W. Tucker in this Oxford book devotes an essay to speculating on the kind of creature this "Pro-Avis" missing link may have been. Pycraft, he says, imagined an arboreal parachuting reptile taking flying leaps from tree to tree. Nopsca developed the idea of a two-legged running reptile waving its arms to increase speed. Mr. Tucker himself favoured the notion of an arboreal reptile with legs adapted for springing, a grasshopper kind of creature. Another scientist, Steiner, helped things on with a theory that wings developed by the edges of reptilian scales fraying out to form feathers. This problem in evolutionist "science" seems suitable for reference to the crossword puzzle fraternity. The fact of the matter is that the extinct giant flying reptiles had wings on quite a different plan from bird's wings. And there is no more evidence of how these wings were evolved than there is of the evolution of bird wings.

Putting these imaginings on one side, let us return to the *Encyclopaedia Britannica* for light on what has been actually observed in the rocks. Professor Julian Huxley in the encyclopaedia's evolution article has given us his word that the fossils provide "complete proof," and the reader is referred to the article on palaeontology to view this proof. There we are told that evolution is evidenced by the Ammonite series of fossils, the Zaphrentis coral series, the Micraster sea-urchin series, and finally the great horse series. None of these series shows one kind of animal changing into another kind. All that is exhibited is minor modification. In the account of the Ammonites the words "it is claimed" appear throughout. Of the Zaphrentis corals, the encyclopaedia says, "It is difficult to select a single case which can be regarded as conclusively established."

The Micraster sea-urchin makes a better effort. Mr. Dewar in his *More Difficulties* (pp. 195-207) deals with him on the same lines as the encyclopaedia but more fully. It seems that Dr. Rowe, the great Micraster authority, sorted out 2,000 of these fossils according to geological age and traced out a continuing change in eighteen different characteristics. The Micraster changed his mouth in the course of ages from a circular to a crescent shape, he grew a lip, altered the lay-out of his spines, and so on. Yet when all was done he was just as much a Micraster sea-urchin as when he began. And there was nothing to show that the last Micraster of the series was any more than a variety fertile with the first. As nobody disputes the occurrence of variations, there is little here to help evolution.

However, the famed horse series of fossils is evolution's chief exhibit. The horse, according to the evolutionists, starts off about the size of a dog, and with five toes on his hind feet and four toes on his forefeet. He then decides that he has overdone evolution and starts shedding surplus toes until he finishes up with one per foot all round. Multi-toed horse fossils are found both in Europe and America. Some evolutionist horse pedigrees contain solely American fossils, some mix American and European fossils, and a less complete series can be made of European fossils only. Mr. Dewar prints in his *More Difficulties* (p. 144) two pedigrees by well-known palaeontologists, one with seven intermediate forms between *Eohippus* and the modern horse, and the other with five. The pedigrees have only two of these intermediate forms in common, and the rest of the ancestry is different.

This horse series, like the sea-urchin series, begins with a horse and ends with a horse. It does not exhibit any other kind of animal turning into a horse. There is nothing written on the fossils to say that the one-toed horse is descended from the many-toed horse. He may be, or he may not. According to the evolutionsts, the one-toed horse appeared later. The anti-evolutionists say these statements are not above suspicion. They quote instances of rocks being arbitrarily re-dated to a later age because one-toed horses have shown up in the fossils in these rocks. Major Wren's *Evolution—Fact or Fiction?* (p. 86) says it is recorded as a well-authenticated fact that Mr. John T. Reid, a mining engineer, found fossil remains of a one-toed horse in a cretaceous formation in a coal-mine in Utah. If this is correct it makes the modern horse very much older than his supposed ancestors.

Mr. Dewar in his *More Difficulties* (pp. 139-148) thinks it possible that the one-toed horse may have been in existence all through. The present writer has no competence to express any opinion on the matter. All he can say is that there are so many admissions by evolutionists of faked or doctored evidence in different directions that a little more is neither here nor there. Everything is supposition. Evolution is supposed to be survival of the fittest, and the writer in turning over an old file of *Nature* noted Major Leonard Darwin saying in an address at the opening of Down House as a memorial to his famous father, that no one had been able to fathom what benefit the horse was supposed to have derived by shedding toes.

Two opinions by well-known scientists on the vexed horse question may be of interest. Deperet, the French palaeontologist, says in his *Transformations of the Animal World* (p. 105) : "The supposed pedigree of the horse is a deceitful delusion, which . . . in no way enlightens us as to the palaeontological origin of the horse." Professor J. Bell Pettigrew, F.R.S., professor of anatomy at St. Andrews, said in his *Design in Nature* (1908, vol. i, p. 217) : "By no means a strong case has been made out for the descent of the horse from a five-toed extinct mammal. Perhaps even less can be said when the teeth of the horse and its supposed ancestor form the fulcra of the argument." It seems unnecessary to say more on the most-paraded item of evolutionist evidence from the fossils.

Of the other vertebrate fossils, the encyclopaedia says : "The equally complete stories of the camel, dogs, and titanotheres have not yet been published and cannot be intelligibly summarised." It is added that "the rhinoceros series is very complex." It is a reasonable inference that if these fossils provided any great evidence in support of evolution it would long ago have been placed on view.

The foregoing constitute the main items in the "complete proof" which evolutionists claim is to be found in the rocks. The anti-evolutionists do not appear irrational when they confess themselves unable to find any grounds therein for believing that fishes have turned into elephants. Mr. Dewar, in his *More Difficulties* remarks after sixty pages of detailed examination of the various evolutionary fossil series, that even if the claims are accepted at face value, the argument amounts to no more than inferring that a man can run 100,000,000 yards in 11,000,000 seconds because he has been timed to run 100 yards in 11 seconds.

Modern scientific literature enlarges on everything which can be made to support evolution, and passes over in silence all that lends no support to the theory. The public thus seldom has its attention directed to the fact that the earliest of all known plants show little signs of evolution. Yet the following statement made many years ago by Sir J. W. Dawson, F.R.S., F.G.S., in his *Geological History of Plants* still holds good : "The old Cambrian and Silurian seas were tenanted with seaweeds not very dissimilar from those of the present time." Present-day evolutionist chronology dates the Cambrian period to about 600,000,000 years ago, and the Silurian to about 400,000,000 years, and *Homo sapiens* to no more than 25,000 to 40,000 years back. The sea-weeds have had longer time than any other plants in which to transform themselves, and there is no transformation. Why ?

The most numerous by far of all animal things, and, according to evolutionists, older by far than reptiles, birds, and mammals, are the insects. Some entomologists estimate that of all animal species insect species account for 80 per cent. The lowest estimates show insect species as considerably over half the total of animal species. In total numbers of individuals, insects must be at least four-fifths of the world's animal population. Pick up the average evolutionist book, and you will find very little about this immense principality of the animal kingdom. What have the fossils to reveal of insects? In the Smithsonian Report for 1931 Mr. R. E. Snodgrass, of the Bureau of Entomology of the United States Department of Agriculture, writes (p. 443) : "The oldest known insects of the geological records are so much like modern insects that palaeontology gives little assistance in a study of insect structures. Probably no other group of animals have so effectively covered their evolutionary tracks as have the insects." According to the current timetable, insects first appeared about 350,000,000 years ago, and among the earliest were the silverfishes and cockroaches we have with us virtually unchanged today. Where is the evolution?

The all-pervasive influence of evolutionist dogma is noticeable in Mr. Snodgrass's statement quoted above. In remarking that palaeontology provides little of no evidence of insect evolution, lip-service is at the same time paid to evolution theory : the insects are said to have "covered their evolutionary tracks." That of which there is no evidence is implicitly assumed. From Mr. Dewar's *More Difficulties* (p. 172) one gathers that scientists endeavour to give an illusion of evolution among insects by assigning different species names to identical insects when appearing in different geological formations.

Darwin and his disciples have never been short of theories to account for the lack of fossil evidence of evolution. Darwin in his *Origin of Species* after significantly heading his first chapter on the subject "The Imperfection of the Geological Record," went on to say that the succession of forms in the rocks was consistent with evolution, and that it would be very wrong to "falsely infer" because the supposed intermediate forms were not there, that they had not existed. "Negative evidence," he affirmed " is worthless." In his introduction Darwin had said there was no scientific advantage in evolutionist belief over creationist belief unless evolution could be established as fact. When he comes to fossils he says what cannot be found must be imagined to have existed.

To account for the total lack of fossils in the Pre-Cambrian rocks Darwin offered the reader a choice of two theories. One was that the continents and oceans had since changed places, and that the missing fossils are now under the sea. This cannot be either proved or disproved until someone finds a way of draining off the oceans and having a look. Alternatively, Darwin suggested that the weight of the later strata might have squashed the supposed Pre-Cambrian fossils out of existence. Mr. Dewar in his *More Difficulties* points out that some

Pre-Cambrian rocks still bear impressions of rain drops and of ripple-marks where water had run over them in the days when they were loose sand. If such things remain, it is odd that all the multitude of supposed fossils should have vanished. Mr. Dewar catalogues five main theories and various sub-theories advanced by later evolutionists to account for the absence of Pre-Cambrian fossils. If one theory is accepted all the rest must be rejected.

Darwin had another theory ready to account for the absence of the intermediate forms between species. He argues that the fossil-iferous beds were deposited during periods in which the land was subsiding, and that in between times were long stationary ages when no fossilisation of animal and plant remains took place. In these stationary periods all the evolutionary transformations of species occurred. Evolution is thus a highly private affair, with everything done off-stage altogether from protoplasm to the Cambrian fossils, and with retirement to the dressing rooms for every subsequent transformation.

The stock cry of evolutionists from Darwin's day to the present time has been "the imperfection of the geological record." It is next door to a miracle, they assert, for any living organism at all ever to become fossilised. Well, it so happens that Mr. Dewar, whose valuable work has been so freely quoted herein, and Mr. G. A. Levett-Yeats, both Fellows of the Zoological Society, went to some trouble to compile statistics about fossilisation. Their figures showed that fossilisation was by no means so miraculous as had been supposed.

Taking existing genera of land mammals (that is, all mammals except bats and aquatic ones) these zoologists found that in Europe, where fossil-hunting had been most intense, these mammals had 100 per cent. representation in fossils. North America came next with 90.14 per cent., followed by South America, 72.09 per cent.; Asia, 70.15 per cent.; Africa, 49.65 per cent.; and Australia, 45.83 per cent. The figures thus indicate that if you look hard enough and long enough you have good prospects of finding fossil remains of all existing land mammals. Mr. Dewar holds that "theoretical considerations indicate that the chances are great that some specimens of every genus having hard parts will become fossilised during the period of the existence of that genus."

The interesting and significant part of the story is that a paper embodying the results of these investigations was offered by Messrs. Dewar and Levett-Yeats to the Zoological Society of London, of which they were both fellows. The paper (subsequently accepted by the Victoria Institute, vol. lxiv, 1932), was rejected on the grounds that its results led to no useful conclusions. The result was certainly not "useful" in helping on evolution's argument that what cannot be found must be imagined to have existed. Mr. Dewar further records that on the leading scientific journal Nature in 1937 publishing an assertion by a correspondent that fossilisation was "almost a miracle," he wrote a short letter giving statistics, and this letter *Nature* refused to publish. Such are the methods by which belief in evolution is maintained and propagated in scientific circles in this dark age of ours. The above facts appear in Mr. Dewar's *More Difficulties*, chapter xvi.

The theory of evolution postulates some enormous transformations in animals. It further requires that every stage in these transformations shall be beneficial to the animal. There is no question of closing down during reconstruction, or even of putting up with present discomfort for future gain. Darwin laid it down in a letter in 1859 that every stage in the change must in itself be of advantage. The eminent Professor Goodrich is similarly quoted in the Oxford *Evolution* book of 1938 (p. 274), and he adds, "it is often difficult to picture the intermediate conditions."

Mr. Dewar goes slightly further than Professor Goodrich. He says it is not "difficult" but "impossible" to imagine some of the required transformations. In his *Challenge to Evolutionists* (pp. 52-57) he defies anyone to picture the conversion, for example, of a land mammal into a whale—which conversion all evolutionists assert took place. Each stage, remember, must make the animal fitter for existence than it was before. The land mammal, says Mr. Dewar, must first be converted into a seal-like creature; it must give up using its hind legs and drag them about behind it until hind legs and tail eventually grow together. Its pelvis must shrink in size, and the portion of the body behind the pelvis must somehow twist round on the front part so that the sides come uppermost and undermost, and the lateral motion of the seal-like stage is converted into the vertical motion of the whale stage.

The whale, moreover, gives birth to its young in the sea and suckles it under water. The mother has to develop muscles enabling her to force milk into the mouth of the young one. She has also to develop a cap round her nipple into which the snout of the young one fits tightly. The young one also requires to have its windpipe prolonged above its gullet to prevent the milk ejected by the mother from entering its lungs. All these modifications have to be effected *before* the young ones are born in the water. There can be no intermediate stages, Mr. Dewar points out, between being suckled in the air and suckled under water. Either sudden miraculous change must be imagined, or equally miraculous prophetic evolution with everything fixed up in advance of the event.

Such are the magical requirements of some evolutionist transformations, of the occurrence of which transformations neither the fossils nor anything else yields the slightest evidence. The imagination is required to take leap after leap. It now remains to view the summit of this monument of human credulity, the amazing collection of balderdash asserted to establish mankind's descent from the beasts of the field.

Chapter V

THE MONKEY-MAN FABLE

DARWIN discreetly waited twelve years after 1859 before applying his evolutionary speculations to human ancestry. His first lieutenant, Huxley, wasted no time in beating about the bush. In his scientific addresses, writings, and lectures to working men and others, Huxley forthwith preached the kinship of man and ape. A large section of the public today believes man's descent to be a scientifically established fact. Actually, this portion of the evolution theory has not so much as a feather to fly with.

Some years ago a London jury awarded a certain peer of the realm £30,000 in damages against the proprietor of a newspaper which had wrongfully described him as of Jewish descent. It would be an excellent thing if those of us who resent the evolutionist libel that our ancestors were apes or other animals were to club together and bring into court all the publishing-houses, professors, etc., proclaiming it. Nothing more would ever be heard of evolution if evolutionists were forced to come to light with evidence that a judge and jury would accept. If they failed to prove their words damages on a much higher scale than in the case quoted could very justly be claimed, for the moral harm done by evolutionist lying is immense, and this particular lie is a vile one.

The evidence in the monkey-man case rests on certain fossilised remains. The first of these to appear, consisting of a skull-cap and some fragments, was found in a cave near Dusseldorf shortly before the *Origin of Species* was published. Numerous other skulls of similar type have since been discovered, and are known as Neanderthal man. The earlier skulls were too incomplete to show the position of the aperture by which the spinal cord entered the brain. It was thus impossible to tell whether Neanderthal man stood upright or not. In conformity with their evolutionist imaginings, scientists for a long time assumed that Neanderthal man was a shaggy, crouching, ape-like creature. A reconstruction of him on these lines was made by the eminent Professor Boule, director of the Museum of Natural History in Paris. Similar models appeared in other museums, and pictures were freely published. A few scientists protested, but they were ignored. In 1929 some more complete remains were discovered at Rome, and showed Neanderthal man to have had an erect human posture and the evolutionist assumptions to have been unfounded (Dewar, *Man*, p. 38). Dr. H. H. Woollard, F.R.S., professor of anatomy in the University of London, in *Science Progress* for July, 1938, describes Neanderthal man as a primitive being, below but nearer to the Australian black than the Australian black is to the modern European.

Rhodesian fossil man is now ranked as of Neanderthal type, and Professor Wood Jones in his *Man's Place among the Mammals* has pointed out how lack of "a little elementary anatomical knowledge" (coupled no doubt with evolutionist dreamings) similarly caused Mr. W. P. Pycraft in a British Museum report in 1928 unwarrantably to turn Rhodesian man into another crouching ape-like being (quoted by Dewar. *Man*, p. 38).

Today the three chief alleged fossil "missing links" between man and ape are Java man, Piltdown man discovered in England, and Peking man. Apart from these is Heidelberg man represented by nothing but a massive fossil jaw-bone described by most anthropologists as essentially human but with some simian characteristics. It may be noted that *Whitaker's Almanack* for 1931 recorded that the remains of Deeming, a notorious Australian murderer of 1892, had been exhumed and were reported by Sir Colin Mackenzie, director of the Australian Institute of Anthropology, to have remarkable simian characteristics. Heidelberg man may thus have had as much or as little connection with the apes as Deeming had.

Java man, otherwise *Pithecanthropus erectus*, or Trinil man, was discovered by Dr. Eugene Dubois. In 1887 Dr. Dubois, then holding a junior position on the staff of Amsterdam University, surprised his colleagues by refusing promotion, and announcing his intention of going out to Java as a Dutch army doctor in order to hunt for evolution's missing link between man and monkey (*vide* Prof. Elliot Smith in *Smithsonian Report*, 1931). In 1894 Dr. Dubois duly returned with his alleged monkey-man and became the lion of the scientific world.

These fossil remains, which have ever since been the subject of controversy, consist of a skull-cap of chimpanzee type, with no forehead and beetling brows; two molar teeth; and a diseased thigh-bone of human type and abut the size of that of a man 5ft. 7in. high. They were discovered by Dr. Dubois near Trinil in central Java, in a part of the bed of the Bangawan River only uncovered in the driest part of the dry season.

The point at issue is whether any ground exists for assuming these remains to have belonged to one and the same individual. The owner of the skull-cap obviously had a head very like that of a chimpanzee, but of exceptional size, for the largest existing ape has a cranial capacity of about 625 c.c., and the scientists figure it out that the Java skull-top indicates a cranial capacity of about 900 c.c. Here it may be noted that an Australian aboriginal had a 1,250 c.c. brain-case, and a modern European averages around 1,400 c.c.

Whoever owned the Java thigh-bone very obviously stood up-right, which no ape does. As for the two teeth, they are generally described as ape-like but unusual. Combine the fragments, and the result is a creature standing erect, with chimpanzee brows and no forehead, a human thigh, and with face, feet, body and arms left to be sketched in according to fancy.

Dr. Dubois' great discovery began with the finding of a tooth in the riverbed in September, 1891. A month later he discovered the skull-cap a yard away. Continuing his explorations of the same locality in September, 1892, he found the thigh-bone 50 feet away from where the skull-cap had been; and also the second tooth, 13 feet away from the thigh-bone and in the direction of the skull-cap. The skull-cap was water-worn and eroded. The teeth and the thigh-bone had their contours clear and sharp, and apparently had not moved since their original deposit.

Dr. Dubois attributed the erosion of the skull-cap to seepage from a cliff on the river-bank. In the report of the Smithsonian Insti-tution of the U.S.A. for 1898 appears the text of an address delivered by Dr. Dubois to the Berlin Anthropological Society in 1896; and in the Smithsonian Report for 1913 is a very full paper on all the re-mains of fossilised man known to that date, compiled by Dr. A. Hrdlicka, curator of the Division of Anthropology of the U.S. National Museum, after a special mission to Europe to examine them for the Smithsonian Institution. From the plans, illustrations, and letter-press in these reports, it appears that the eroded Java skull-cap was up-stream from the uneroded teeth and thigh-bone. But no doubt water would run uphill to help on evolution.

In his Berlin address on his discoveries, Dr. Dubois stated that "associated with these bones" he had found fossil remains of Steg-odon (an extinct elephant) and of a small deer, and "further away" remains of buffalo, antelope, ox, pig, rhinoceros, and hyaena. Sir Arthur Keith in his *Antiquity of Man* says that altogether Dr. Dubois removed from this spot in the bed of the Bangawan River between 1891 and 1894 fossils of twenty-seven different kinds of mammals. A German expedition under Madame Selenka also spent two years from 1906 making much more extensive explorations in the same spot and unearthed an immense quantity of miscellaneous fossils, but got no traces of monkey-men.

It thus appears that the scanty fragments constituting Java man were sorted out by their discoverer from a pretty complete Noah's Ark bone-heap. This discoverer, moreover, is stated to have sacrificed a superior position in Amsterdam for an inferior position in Java for the express purpose of hunting for the monkey-man missing link predicted by evolution. Most of us in such case would be prone to view in the most hopeful light whatever oddments our digging in tropical riverbeds might produce. We would have our monkey-man or perish in the attempt.

Dr. Dubois told the Berlin savants in 1896 that "in other situations of the same stratum" he had found fossil remains of a gigantic scaly ant-eater and of hippopotamus. He did not tell them that in these "other situations" there had been discovered in 1889 at Wadjak about fifty miles from Trinil, a fossil human skull of Australian black type, and that he himself in 1890 had there unearthed a second fossil skull of the same type. These interesting facts Dr. Dubois strangely withheld from publication until 1920, fully thirty years on. According to *Nature* for January 6, 1921, he then casually disclosed them following on discussion of a find of fossil man at Talgai, Queensland. Possibly Dr. Dubois had good reasons for keeping quiet for so long. At the same time it requires no great powers of penetration to perceive that the cause of evolution might not have been helped by revealing true man as in existence along with his supposed monkey-man ancestor. As Sir Arthur Keith remarks, to have put all the articles on the table simultaneously would have provided the learned with more than they could digest—perhaps with more even than they could swallow.

Java man—still evolution's chief mainstay—appears throughout his career to have been shrouded from the gaze of profane eyes. One finds Dr. Hrdlicka writing thus in the *Report of the Smithsonian Institution* for 1913 : "All that has thus far been furnished to the scientific world is a cast of the skull-cap, the commercial replicas of which yield different measurements from those reported taken of the original, and several not thoroughly satisfactory illustrations : no reproductions can be had of the femur and the teeth, and not only the study, but even a view of the originals, which are still in the possession of their discoverer, are denied to scientific men." Dr. Hrdlicka, official emissary of the great Smithsonian Institution, presided over by the President and Chief Justice of the United States, was refused permission even to inspect the originals. He described the position as "anomalous."

Since 1936 two incomplete skulls and some skull fragments, similar to the Java man skull-cap, have been discovered at Sangiran in Java by Dr. G. H. R. von Koenigswald, as recorded in *Nature* of December 2, 1939. The most complete of these gives a cranial capacity of 835 c.c., according to its discoverer, as against the 900 c.c. estimated for the Dubois Java man. No human-like thigh-bones or other skeletal parts had been discovered up to the last report seen by the present writer.

There has never been the least agreement among scientific men that Dr. Dubois was justified in assuming his skull-cap and thigh-bone to belong to the same individual. In his Berlin paper of 1896 Dr. Dubois tabulated the opinions of about a score of leading scientific men on the remains, showing the utmost variance. Finally, to cap all, Dr. Dubois himself in 1938 announced that after prolonged study of anthropological textbooks, of the *Pithecanthropus* bones, and *"of other material from the same provenance in his possession, for the most part not previously published,"* he was of opinion that "we are here concerned with a gigantic gibbon."

In making this announcement in its issue of February 26, 1938, *Nature* flatly refused to yield up Java man as evolution's prize exhibit. Dr. Dubois' new conclusions about his fossils, it said, had been received "with respect, but not with general acceptance; and in the light of the new evidence must be regarded as definitely disproved." This leading scientific journal then asserted : *"Pithecanthropus* now stands within the line of human descent, if only as a pre-hominoid."

It thus appears that the editor of *Nature* is a better authority on Java man than its discoverer. Netherthless, whatever *Pithecanthropus* may have been, it is quite impossible for him to have been ancestral to man, as we shall see later. *Nature's* dogmatic statements, moreover, are extremely rash in view of Dr. Dubois' announcement that even after half a century he still has some cards up his sleeve in unpublicised Trinil fossils. Java man would probably never have been heard of, had Dr. Dubois in 1894 placed his two Wadjak human skulls on the table alongside his Trinil chimpanzee skull-cap and human thigh-bone. Java man is hardly likely to be abandoned by his discoverer without the very best of reasons.

Piltdown man, otherwise *Eoanthropus*, or Dawn man, is the next item on the monkey-man list. He hails from a fossil-bearing stratum six inches thick, near the bottom of a small gravel pit, four feet deep, used for metalling a by-road on the Sussex Downs, eight miles north of Lewes. Piltdown man consists of nine small fragments of skull-bone, and rather less than half of a chimpanzee-like jaw bone. There was not much of him altogether, and he was discovered in sections over a considerable period of years by Mr. Charles Dawson, solicitor of Lewes, an amateur fossil-hunter. Toward the end the assistance was secured of Dr. (later Sir) A. Smith Woodward, of the British Museum staff and soon afterwards president of the Geological Society.

This jig-saw puzzle was laboriously fitted together—so far as it would fit—to form part of the top and back of a skull. The vacant spaces were filled with plaster of paris, with forehead and facial bones duly modelled in plaster. Finally, the jaw-bone, with its missing three-fifths also completed in plaster, was neatly hung on in front. The resulting monkey-man was then exhibited to a crowded and sensational meeting of the Geological Society in London on December 18, 1912.

Piltdown man as first presented was announced to have a cranial capacity of 1,070 c.c., which puts him ahead of Java man's 900 c.c., but well below the Australian blacks' 1,250 c.c. He has been several times reconstructed—apparently with still more generous assistance from the plaster-pot—as he is nowadays quoted as measuring 1,300 c.c. in cranial capacity.

The chief of the numerous points at issue in this highly conjectural item of evolutionist evidence is whether the chimpanzee jaw-bone ever had any real connection with the human skull fragments. It is a debatable point, and there is the utmost diversity of opinion in scientific circles on it. There is reason for doubt, for along with the skull fragments and jaw-bone there were extracted from this six-inch wide fossil-bearing stratum the following things : A tooth of a mastodon, a tooth of a Stegedon (an extinct elephant previously unknown in Western Europe), two teeth of a hippopotamus, two teeth of a beaver, the femur of an elephant shaped up for use as a tool, and finally some flint implements. There would appear to be nothing wildly incredible in a real chimpanzee having contributed part of a jaw-bone to this miscellaneous zoological collection.

Piltdown man got the late Mr. Charles Dawson a monument, and helped Dr. Smith Woodward on to the presidency of the Geological Society in 1914 and a knighthood later on. His claims are by no means universally accepted, for one finds Professor Sir Grafton Elliot Smith stating in 1931 of this evolutionist exhibit : "Even today many Continental anthroplogists refuse even to refer to it in treatises on fossil man, or when they do, brush it aside as so doubtful that it is best to ignore it" (Nature, June 27, 1931).*

Peking man, the final exhibit of the series, next presents himself to our view. He hails from the floor of a cave in a disused limestone quarry, thirty-five miles southwest of Peking in China. In the Smithsonian Report for 1931 Professor Elliot Smith stated that Dr. Davidson Black, on learning of the discovery of a peculiar fossil tooth somewhere near Peking, went out to China to join the Chinese Geological Survey in the hope of finding a fossil monkey-man. In 1929 Dr. Black made his first discovery of an incomplete skull. A considerable number of skulls have since been found, one or more complete enough to show the nose as broad and flat; and the cranial capacity of the skulls is put at from 1,000 c.c. to 1,100 c.c.

These remains so closely resembled the chimpanzee-like Java skull-cap that there was long argument whether Peking man was sufficiently distinct from Java man to be allotted a scientific name of

---

* In an article in the Times of Nov. 21st 1953 a British Museum correspondent admitted the fraudulency of Piltdown Man. (Editor)

his own. Eventually he was christened *Sinanthropus*, though some scientists affirmed that there was next to nothing to distinguish him from *Pithecanthropus erectus* of Java. In *Nature* of December 2, 1939, it was recorded that seven thigh-bones of Peking man had been discovered, mostly incomplete shafts, and according to the descriptive matter they lacked the human characteristics of the Java thigh-bone. Incidentally, it may be noted that along with Peking man there were also found in the cave floor remains of over fifty types of mammals, as well as fossil frogs, snakes, turtles, and birds. Up to 1930 no less than 1,475 cases of fossil bones were removed from the site (vide Prof. G. B. Barbour at the British Association, *Nature*, September 27, 1930).

Java man got his semi-human attributes by assumption that a human thigh-bone had belonged to a chimpanzee skull-cap. Peking man is Java man over again, but without any human thigh-bone. In view of the condition of uncritical credulity induced in the scientific mind by the evolution theory, the layman must be pardoned for wondering if the position is that Peking man climbs up to semi-human status on Java man's knees—or Java man's thigh-bone, to be precise. If such is the case, one can understand the total havoc which would be wrought in mankind's evolutionary ancestry were Dr. Dubois permitted to fling Java man to the wolves as a mere ape and nothing more. Not only would evolution's No. 1 exhibit vanish, but down with it would crash Peking man also. The sole remaining monkey-man would then be Piltdown man. And when the plaster of paris is removed how much is left of Piltdown man? Nothing but a few fragments of bone which look as if they might all be packed up inside a breakfast cup. It may be that the editor of *Nature* scented this impending tragedy when he so flatly refused to part with Java man on any consideration whatsoever—not for all the Dr. Dubois in the world.

We now come to the final point. No matter just what Java man, Piltdown man, and Peking man may have been, it is quite impossible for them to have been ancestors of man. The reason is that full-fledged man, *Homo sapiens*, was already in existence, cooking his breakfast, making his tools, and going about his daily business when evolution's alleged missing links appeared.

Mr. Dewar, F.Z.S., in his *More Difficulties of the Evolution Theory* (p. 93) points out that fossils of men of modern type have been discovered in deposits "certainly at least as old as, probably older than" those containing Java man, Peking man, etc. He enumerates the Castenedolo, the Olmo, and the Calaveras fossil skulls, the Oldoway and Clichy skeletons, and the Abbeville, the Foxhall, the Kanam and Kanjera fossil jaws. Mr. Dewar is an anti-evolutionist, and anti-evolutionists are beyond the pale in well-conducted scientific circles. We therefore lay his statements aside.

We turn instead to the leading British scientific quarterly review *Science Progress* issued by the highly respectable publishing house of John Murray. In the number for July, 1938, we find there an article on "The Antiquity of Recent Man" by Professor H. H. Woollard, F.R.S., professor of anatomy at University College, London. Needless to say, Professor Woollard is an evolutionist. No anti-evolutionist would for one moment be permitted to occupy the post he holds. However, Professor Woollard is an unusually candid evolutionist.

In his article he reviews the various fossil men. He thinks "there cannot be absolute certainty" that the two bones, plus two teeth, constituting Java man belonged to the same individual. The thighbone is "indubitably human," and the skull-cap "recalls in many ways the form of the acrobatic gibbon." Java man and Peking man form "one type." The Heidelberg jaw is "essentially human," but "the resemblance to the anthropoid jaw is especially close." The Piltdown skull fragments differ only from a modern skull in being unusually thick, and the jaw "resembles most closely that of a chimpanzee and looks incongruous with the skull." In view of later discoveries of fossil man in England, Professor Woollard rejects the jaw as in no way connected with the skull fragments. This leaves Piltdown man just a plain human being.

As to the age of the various remains, Professor Woollard puts Java man and Peking man as contemporaries in the Lower Pleistocene; Heidelberg man in the Middle Pleistocene; Piltdown man in the Lower Pleistocene, or even earlier; and Neanderthal man in the Upper Pleistocene. As previously stated, Professor Woollard describes Neanderthal man as a more primitive human being than the Australian black, but nearer to him than he is to the modern European.

Professor Woollard remarks that Java man, Peking man, and Neanderthal man form a series rising in cranial capacity, and are regarded by palaeontologists as forming a sequence in the emergence of man from the lower animals. He adds : "The difficulty in feeling content with this view arises because in sharp contrast with these fossil types others have been discovered which are in no way different from modern man, and which are as old, or even older, than those just described." Professor Woollard remarks that, "obviously people living contemporaneously cannot be ancestors to one another."

"The other aspect of the matter, "he continues, "is illustrated by a series of fossils which have been found in various parts of the world, but curiously with quite uncommon frequency in England. A series of very ancient fossils has been found which attest the fact that the modern Englishman, so far as his anatomy goes, extends backwards into the past to a time when in other countries man was distinguishable with difficulty from the ape."

The English skulls referred to by Professor Woollard are the

Swanscombe skull discovered in 1937; the skull discovered in 1925, in excavating for foundations for Lloyds Bank in Leadenhall Street, London; and a third fossil skull found at Bury St. Edmunds. All these he says, date at least to the Early Pleistocene.

Professor Woollard's own evolutionist view is that "man started abruptly, and that in the ancestral stock there was a period of great instability and change, and by mutations many new types were evolved." In other words, you go to bed one night as a chimpanzee (or whatever animal is preferred), and wake up next morning as full-fledged man, with powers of speech, a taste for music, and a faculty for mathematics. Science will swallow anything to dodge away from the idea of God Almighty having had anything to do with man's appearance upon the earth. Needless to say, Professor Woollard adduces no evidence in support of a sudden magical transformation of animals into men.

The degraded condition of much present-day biological science is pretty evident from some of Professor Woollard's candid statements. Consider, for example, the implications of the following :

"If two fossil men are found on the same geological level, and one has a large brain and the other a small brain, invariably it will be shown on a genealogical map that the man with the small brain emerged much earlier from the common stock than one with the big brain. This, of course, is an anatomical inference drawn so because of preconceptions that the evolutionary process must proceed by gradations. It is not founded upon any knowledge got from palaeontological evidence."

And again : "The discovery that recent man has a vast antiquity, in fact greater than any other variety, most anatomists have always tried to get round or minimise by making all hominoid fossils carry pithecoid features which are absent in present-day man."

These are plain and open admissions by a front-rank evolutionist that evolution is kept going by faked facts and doctored evidence.

We have now gone over the evolutionist "evidence"—if such it can be called—of man's alleged animal descent. Mr. Dewar states no more than plain fact when he says that despite a search extending over the greater part of a century, "it is not possible to point to any fossil and say of it : the species represented by that fossil, while not human, is ancestral to man" (*More Difficulties*, pp. 93-94).

Having looked over the actual facts in some detail, it is instructive to note the kind of statement to be found in book after book on the shelves of the public libraries in English-speaking countries today. Typical of what is palmed off on an unsuspecting public as established fact, is the following, from *Man, the Slave and Master* by Dr. Mark Graubard, published by Dent and Sons, London, in 1939, after first appearing in the United States :

"The oldest fossils pertaining to man, almost a true missing link, is the ape-man of Java, with a brain intermediate in size between ape and man, yet walking upright like a man, as his skeletal structure indicated beyond a doubt. There is also the Piltdown or dawn-man with large canines and small forehead and ape-like jaw. And finally we have the Peking man, definitely outside the genus homo, but more advanced·than the ape. The oldest distinctly human fossil is probably that found near Heidelberg and called the Heidelberg man . . . The ape-man of Java and the Piltdown and Peking men all existed about a million years ago . . . The Heidelberg man roamed the earth about half a million years ago."

A suburban grocer selling under-weight butter or adulterated foodstuff is hauled before the courts and punished if detected, but it is nobody's business that reckless evolutionist rubbish, without a rag of fact to support it, is ground out wholesale on the printing presses to poison the public mind. To their everlasting honour there are a number of scientists who have refused to countenance the monkey-man fabrication.

Here is what Professor Wassmann says in his *Modern Biology:* "It is nothing short of an outrage upon truth to represent scanty remains, the origin of which is so uncertain as that of *Pithecanthropus*, as absolute proof of the descent of man from beasts in order to deceive the general public."

Dr. Clark Wissler, Curator-in-Chief of the Anthropological Section of the American Museum of Natural History, said in the *New York American* of April 2, 1918 : "Man, like the horse, or the elephant, just happened anyhow . . . Man came out of a blue sky so far as we have been able to delve back."

Professor W. Branca, of Berlin, says in *Fossil Man:* ' Palaeontology tells us nothing on the subject, it knows of no ancestors of man." Professor J. Reinke, of Kiel University, says in his *Monism and its Supporters:* "We are merely having dust thrown in our eyes when we read in a widely circulated book the following words : 'That man is immediately descended from apes, and more remotely from a long line of the lower vertebrates, remains established as an indubitable historic fact, fraught with important consequences.' . . . The only statement, consistent with her dignity, that science can make, is to say she knows nothing about the origin of man."*

Mr. Douglas Dewar says in his *Man, a Special Creation:* "The way in which the public is deluded by complete pictures of man's

---

*These four statements are quoted from Major E. C. Wren's *Evolution: Fact or Fiction?* p. 68.

supposed ancestors, based on a jaw or a piece of a skull or even a tooth, is scandalous. The public has no idea that these pictures are pure figments of the imagination."

The extent to which evolutionist imaginings can dominate the scientific mind was illustrated by an incident in the United States in 1922, as quoted in Major Wren's *Evolution—Fact or Fiction?* A single molar tooth was found in a Pliocene deposit in Nebraska. The great palaeontologist, Professor H. F. Osborn, then president of the American Museum of Natural History, described it as belonging to an early type of monkey-man which he duly christened *Hesperopithecus*. At this date, Mr. William Jennings Bryan was denouncing evolution, and Professor Osborn made the discovery of *Heseropithecus* the occasion for the following rebuke to Mr. Bryan: "The earth spoke to Bryan from his own State of Nebraska. The *Hesperopithecus* tooth is like the still small voice, its sound is by no means easy to hear . . . this little tooth speaks volumes of truth"—of man's animal descent.

In England the eminent anthropologist, Professor Sir Grafton Elliot Smith, induced the *Illustrated London News* to publish an article on this ancestor of humanity, illustrated by drawings of *Hesperopithecus* and his spouse—all on the strength of one small tooth. Presently, it was established that the tooth was that of a peccary, a kind of pig, and *Hesperopithecus* disappeared from view. In its 1929 edition the *Encyclopaedia Britannica* felt it necessary to make reference to this lost asset of evolution, but it wrapped up the horrid truth as well as it could by disclosing no more than that the tooth was eventually found to belong to "a being of another order"—which was one way of spelling "pig" in twenty letters.

In 1925 the State of Tennessee passed a law forbidding the teaching of evolution in its schools, and Mr. Bryan, just before his death, successfully appeared as chief anti-evolutionist counsel in a test case at Dayton, which was given wide publicity, with universal newspaper ridiculing of him. The case was promoted and financed on the evolutionist side by the American Civil Liberties Union, and it is not inappropriate to note that six years later a United States Congressional Committee on communist propaganda said of this body, "fully ninety per cent. of its efforts are on behalf of communists who have come into conflict with the law" (House of Representatives' Report No. 2290, 1931, p. 56). Allusion is made to this Tennessee case in a passage in Mr. R. C. Macfie's *Theology of Evolution*, published in 1933. He says:

"So long as the question is as open as at present, it is scandalous that children and students should be taught as a proven fact that their ancestors were apes, and should be shown abominable pictures of primitive man as a shaggy ape-like creature with a low forehead, receding chin, bowed back, and bent legs. Such science is a disgrace to the spirit of science and a crime against humanity, fit only for the yellow press, and the Catholics and Daytonians deserve honour for declining to accept a totally unproven hypothesis."

The gospel of man's animal descent is the crown of the evolution theory, and to it all the rest leads. The evidential standards of modern evolutionist science represent probably the lowest point in intellectual degeneration reached by civilised man in the past two thousand years. All is wildest assumption and limitless credulity, and with no other end in view than to arrive, by hook or by crook, at the most debased view of human origins which the mind of man is capable of conceiving.

Chapter VI

HOW EVOLUTION WAS BORN

A SIGNIFICANT fact about evolution is that the central idea of the Darwinian theory is not based upon anything observed in nature. Darwin states that he opened his first notebook for facts in relation to the origin of the species in July, 1837, at which time he was twenty-eight years old. He had observed affinities between living and extinct species in his work as a naturalist during the voyage of the Beagle. He had also noticed how man had produced varieties of domesticated plants and animals by selection in breeding. These observations had led to belief in evolution. "But," he relates, "how selection could be applied to organisms in a state of nature remained for some time a mystery to me" (*Life and Letters*, i, 83).

Illumination came in October, 1838. It came by Darwin reading the famous *Essay on Population* of the Rev. Thomas Malthus, with its gospel of an intense struggle for existence in consequence of living things increasing faster than food supply increased. Darwin's own observation in his work as a professional naturalist had not impressed any such idea on his mind as the outstanding fact in nature. The reader will do well to reflect whether this supposed eternal remorseless struggle for existence is a thing which drives itself into his own consciousness in his wanderings abroad. Does he witness it in his garden, about the countryside, in the wilderness, in the air, in the waters? Is he presented with a view of a world with every cranny bursting to suffocation with life, and with the surplus production dying from starvation or trampled underfoot? Is this evolutionist idea of "Nature red in tooth and claw," with the mangled remains of the unfit strewn in every direction, something real existing for all mankind to see, or is it a dream—a scientific nightmare?

Darwin certainly had no idea of this fierce struggle for existence as a factor in evolution until he read Malthus. Having adopted the idea as the foundation of his theory, he spent the next twenty years collecting facts to support it. This work involved correspondence with naturalists and others in all parts of the world. Among Darwin's correspondents was a young naturalist named Alfred Russell Wallace. In June, 1858, Darwin received from Wallace, then on an expedition to the East Indies, a paper which Wallace said he would like to have read before a scientific society, and he asked Darwin to arrange this. Darwin looked at the paper, and was prostrated with mortification to find that it set forth exactly his own idea of evolution by natural selection, or survival of the fittest in the struggle for existence.

The question thus arises where Alfred Russell Wallace got his idea of the struggle for existence. How close it was to Darwin's may be gathered from the letter which Darwin in his anguish wrote to his friend Sir Charles Lyell, the eminent geologist : "Your words have come true with a vengeance—that I should be forestalled... if Wallace had read my MS. sketch written out in 1842 he could not have made a better short abstract ! Even his terms now stand as heads of my chapters. .. So all my originality, whatever it may amount to, will be smashed. .. " (*Life and Letters*, ii, 116). Darwin had guarded his great idea very closely, for in a letter to Lyell a week later all he seemed able to think of as proof of his priority over Wallace was that his intimate friend Hooker, the botanist, had seen his earlier sketch, and that more recently he had given Professor Asa Gray in America an extremely vague and guarded outline of his ideas. As history records, this matter of priority was adjusted by Lyell and Hooker arranging for the reading of a paper by Darwin along with Wallace's paper at the Linnæan Society on July 1, 1858. Evolution thus saw the light with Britain's leading geologist and leading botanist standing as sponsors with explanatory remarks.

Wallace had arrived at exactly the same idea as Darwin, and he had reached it in exactly the same way. He had not observed an intense struggle for existence as a fact of nature. All that had happened—as Wallace relates in his autobiography and elsewhere—was that he had chanced to read Malthus on *Population*, and then sat down and dashed off an essay on evolution by survival of the fittest in the struggle for existence. After publication of the *Origin of Species* it came out that Dr. W. C. Wells had read a paper to the Royal Society in 1813 on evolution by natural selection, and that in 1831 Mr. Patrick Matthew had embodied the idea in the appendix to his book on *Naval Timber and Aboriculture*. It does not appear where these writers found their inspiration. There would be nothing wildly improbable in it coming from the same source, for Malthus published his book in 1798 and it was soon attracting the widest attention.

We therefore turn to the famous *Essay on the Principle of Population* as the fountainhead of the mighty river of evolution. At last, surely, we shall find it here related just where in nature the Rev. Thomas Malthus witnessed this remorseless struggle for existence. Again we are doomed to disappointment. Before we finish perusing the first two pages of his first chapter we find Malthus in a footnote referring us to Benjamin Franklin's *Miscellany*, page 9, as the place to learn about the struggle for existence. The present writer has not had access to this last-named book. He is thus unaware whether Benjamin Franklin saw anything for himself, or whether he in turn derived the idea of living things increasing faster than food supply from yet another book. He notes, however, a statement in Buckle's *History of Civilisation* (World's Classics edition, ii, 247) that it was Voltaire who first threw out this foundation idea for so many theories. As Franklin was United States envoy and minister in France in Voltaire's day, and according to his biographers was intimate with Voltaire, it may be that he picked up the notion there. Our search for evolution's origins thus leads back once more to the peculiar brand of philosophy on which the French Revolution was incubated.

Malthus opens his famous book by stating that his object is "to investigate the causes which have hitherto impeded the progress of mankind towards happiness," and in particular "the constant tendency in all animated life to increase beyond the nourishment prepared for it." Malthus then adds :

"It is observed by Dr. Franklin that there is no bound to the prolific nature of plants or animals but what is made of their crowding and interfering with each other's means of subsistence. Were the face of the earth, he says, vacant of other plants, it might be gradually sowed and overspread with one kind only, as for instance with fennel; and were it empty of other inhabitants, it might in a few ages be replenished from one nation only, as for instance with Englishmen. This is incontrovertibly true. . ."

An Australian writer, Mr. O. C. Beale, in his *National Decay* (1910, p. 33) remarked that so far from being incontrovertible truth, the spread of fennel quoted by Malthus was a strange and venturesome untruth. No botanist, he said, would admit the statement as being true for fennel or any other plant. Nor had anybody knowledge that Englishmen could survive through ages all over India or throughout Africa.

Mr. Beale noted, to begin with, that this supposed profound truth of Malthus—in reality the foundation of the most colossal errors of all time—is remarkable for its laxities of expression. "Causes," for instance, cannot impede progress : only impediments impede. Is it so sure that there is a "progress of mankind towards happiness ?" In any case this abstract and intangible phrase is no basis for a proposition of importance. As for "the constant tendency in all animated life to increase beyond the nourishment prepared for it," we may assume that by "prepared" Malthus means "available," as otherwise the preparation would be in default. Mr. Beale remarks that Malthus was asked by someone what he meant by a "tendency" which nowhere had the effect ascribed to it. The question remained unanswered. Where was the evidence that mankind, for instance, had ever pressed upon the planet's limits of production ?

The social philosophy of the Rev. T. Malthus was expressed in its most naked form in a paragraph in his first edition, but which he

discreetly suppressed in later English editions. As translated back by
Mr. Beale from the French edition, it reads as follows :

"A man born into a world already occupied, if his family can no
longer keep him, or if society cannot utilise his work, has not the least
right whatever to claim any share of food, and he is already one too
many upon the earth. At the great banquet of Nature there is no cover
laid for him. Nature commands him to go and she is not long in
putting this order herself into execution."

Malthus, writing at a time when the British Isles had a population
of 12 millions as compared with their present 47 millions, urged immediate restriction of population as the way to social betterment.
He preached voluntary limitation of families, abolition of outdoor
poor relief, and the herding of the poor into workhouses with hard
fare and husband and wife separated so as to check breeding. These
eminently Christian ideas were adopted in part in the British Poor
Law of 1834.

The name of Malthus is most prominently associated with the
modern practice of birth restriction. "The infant," said Malthus, "is
of comparatively little value to society, as others will undoubtedly
supply its place. Its principal value is on account of its being the object
of one of the most delightful passions in human nature—parental
affection." France was the first country to adopt the Malthusian
doctrine of baby restriction, and France has been the first great
European nation to go under. In Britain it was preached with immense
success by John Stuart Mill, George Jacob Holyoake, Charles Brad-
laugh and Annie Besant, and the birthrate of the whole British race
has now fallen below the point necessary to maintain population.

From Malthus also was derived in large part the inspiration of
David Ricardo's *Principles of Political Economy*, as Ricardo himself
admitted therein. The Ricardian doctrine of every man for himself and
devil take the hindmost coloured all nineteenth century economic
thought : and, via Ricardo, the Malthusian gospel of the struggle for
existence was transformed by Marx into the class war of his revolutionary socialism. Malthus thus helped to provide bloodthirsty Bolshevism
with its theoretical outfit.

Finally, through Charles Darwin the same Malthusian teaching
became the foundation of the theory of organic evolution, itself in
turn the sheet anchor of materialism and atheism; and with a large
percentage of its adherents attached also to the birth-control and
Bolshevik fruits of the same tree.

Such is the enormous fabrication of error built upon one fantastic
untruth. No one has ever seen this imagined, continuous intense
struggle for existence with all living things pressing to the limits of food
supply, and with mass starvation of surplus population. The whole
thing is a dream. All animals have to bestir themselves to get their

breakfast it is true : but that is all there is to it. Even the orthodox evolutionists today are unable to discover Darwin's imagined struggle for
existence. "It is the doctrine of Malthus applied with manifold force to
the whole animal and vegetable kingdoms," said Darwin in his *Origin
of Species*. Here is what a modern specialist has to report. The extract
is from the Oxford book on *Evolution* of 1938, to which reference has
been made in previous chapters. The writer of the essay quoted is
Mr. Charles Elton, director of the bureau of animal populations at
Oxford University. He says :

"A first impression might be that every niche has long ago been
filled with plants and animals dependent on plants, that the habitats
are full to bursting point with life. . . This concept fits plant life fairly
well, but it is not true of animals. It is obvious to any naturalist that
the total quantity of animal life in any place is an extremely small
proportion of the quantity of plant life. This general observation has
been amply confirmed by all recent studies of the biomass of animal
species or animal communities. For example, the bird life on an acre
of rich farm land with trees and hedges and grass and crops may be
only a few kilograms in weight. The animal life is widespread, it
has, so to speak, staked out its numerous claims, but has seldom succeeded in exploiting them to the full. Only in certain inter-tidal communities of the sea do we feel that animals have reached the limits of
the space that will hold them. Even here it is, cosmically speaking,
a tiny film of life" (pp. 129-130).

"From this situation we may conclude that, on the whole, animal
numbers seldom grow to the ultimate limit set by food-supply, and
not often (except in some parts of the sea) to the limits of available
space. This conclusion is also supported by the general experience
of naturalists, that mass starvation of herbivorous animals is a comparatively rare event in nature. . ." (p. 130).

"Fluctuations occur in every group of animals and in every habi-
tat that has been investigated . . . Although the amplitude of fluctuations is often very great, . . .two things that we might expect do not
often happen. The first, complete destruction of vegetation by herbivorous animals, has already been mentioned. The second is complete
destruction over any wide area of either predators or prey" (p. 131).

Thus this specialist in charge of a bureau set up to investigate the
matter of animal numbers can find no sign whatever of the Malthus's
imagined universal tendency of all living things to multiply and increase beyond the available food supply. The whole Darwinian theory
of evolution, basis of all modern biological science and proclaimed on
every hand as established fact, rests on something that never was on
land or sea.

"Only in certain inter-tidal communities of the sea," says Mr.
Elton, "do we feel that animals have reached the limits of the space

that will hold them." We all know these spaces on certain beaches where at low-water we find a mass of pools and living things. What are the commonest objects meeting the eye in such spots ? The answer is, mussels and seaweed. If there is a struggle for existence, mussels and seaweed are thus in the very mid-centre and vortex of it. Evolution should here be proceeding at top-speed. What are the actual facts? The eminent Professor J. Ritchie in his presidential address to the zoological section of the British Association in August, 1939 handed down the latest bulletin about mussel evolution. He said : "The edible mussel (*Cardium edule*) has retained its specific characters for two million years or more, its genus in a wide sense lived 160 million years ago in the Trias" (*Nature*, Sept. 23, 1939). In other words, so far from evolving into something else, the mussel, packed up twenty to the dozen on his perch, has not changed a whisker in two million years, and was a perfectly good mussel 160 million years ago. As for seaweed, we have already noted the statement by Sir J. W. Dawson, that the seaweeds of today are "not very dissimilar" from those of the Cambrian and Silurian seas—dating back 600 to 400 million years ago, according to the same fashionable evolutionist chronology. Where is the fabled evolution ?

The great evolution theory falls down flat at the very first hurdle, the supposed struggle for existence. This struggle for existence is affirmed to result in Natural Selection. Very well, we get our evolutionist steed up on his feet and start off again. Down he falls once more. "No recognised case of Natural Selection really selecting has been observed," states Professor Vernon Kellogg (vide Major Wren's *Evolution*, p. 91). That is to say, all sorts of individual variations occur in organisms, but no instance at all is known of individuals with favourable variations supplanting individuals with unfavourable variations.

Let us persevere once more. Natural Selection, according to evolutionist doctrine, results in survival of the fittest. Once more evolution fails to make the jump. "The non-utility of specific characters is the point on which Natural Selection as a theory of the origin of species is believed to fail," says Professor D. H. Scott (*Extinct Plants*, p. 22), and many other scientists say the same thing. Species are just different in some way from the next nearest species, but no particular usefulness can be discovered in the characteristics distinguishing the one species from the other.

The actual fact of the matter thus appears to be that there is no intense struggle for existence ; there is no Natual Selection ; and there is no onward and upward evolutionary progress of the more fit replacing the less fit. Every part of the theory of evolution crumbles away to nothing when examined. These things are the foundations of the whole edifice. In earlier pages we have noted the total absence of evidence that living things can break through their breeding limits.

No such thing has ever been observed, and every attempt to produce it by experiment has failed. Eighty years of fossil-hunting has totally failed to produce the intermediate forms between species which evolution says existed. The imagined embryological evidence of evolution is nowadays rejected by evolutionists as amounting to nothing at all. The vestigial organs, the "plain inutility" of which, to use Darwin's words, showed them to be survivals from past forms of being, are all the time turning out to be useful and not useless : they thus steadily and increasingly vanish as evidence of evolution. From top to bottom there is nothing whatever that will hang together. Evolution is the biggest scientific fraud of all time.

We have noted how Darwin spent twenty-one years collecting matter to support his theory of evolution before he published the *Origin of Species* at the end of 1859. His letters show quite clearly that even at the end, on the very eve of publication, Darwin had not succeeded in convincing himself. On November 23, 1859, he wrote to Lyell, ". . . Often a cold shudder runs through me, and I have asked myself whether I may not have devoted my life to a phantasy" (*Life and Letters*, ii, 229). Two days later he wrote to Huxley. "I had awful misgivings, and thought perhaps I had deluded myself as so many have done" (*ibid*, ii, 232). That was all the actual belief in his theory Darwin was able to muster up with advance copies of his book in the hands of his friends. He saw and knew in his heart that he had produced nothing but a patchwork of incoherent guesses. No man who feels the firm ground of truth beneath his feet writes in such a strain as Darwin wrote in these letters.

Beside his main theory Darwin introduced various subsidiary theories into his books. One such, for instance, that runs through the *Origin of Species* is that continental animals and plants are more highly evolved by stress of competition than are insular ones; and that continental productions introduced into an island will supplant and replace the less improved native productions. This notion got into Darwin's head apparently at the end of 1858, for he then mentions it in a letter to Hooker, saying, "See how all the productions of New Zealand yield to those of Europe. I dare say you will think all this is utter bosh, but I believe it to be solid truth" (*Life and Letters*, ii, 143).

In his *Naturalised Animals and Plants of New Zealand* Mr. G. M. Thomson relates how in the early years of settlement introduced plants and animals flourished exceedingly, water-cress, for example, for a few years growing to 14 feet in length, and then later subsiding to normal size again. Mr. Thomson tells how as an ardent evolutionist he waited to see the introduced vegetation replace and exterminate the New Zealand bush. He states that he waited in vain : "The opinion of all botanists in New Zealend today is that when the direct, or—to a large extent—the indirect influence of man is eliminated, the native vegetation can always hold its own against the introduced" (pp. 527-8). He says also, "The same probably holds good to some extent with animal life, only the problem is more difficult to follow out."

Darwin's son, Sir Francis Darwin, as editor of his father's *Life and Letters* said of the paternal propensity for theory-building that "it was as though he were charged with a theorising power ready to flow into any channel on the slightest disturbance" (i, 149). Darwin's elder brother Erasmus wrote back after reading his copy of the *Origin:* "The *a priori* reasoning is so entirely satisfactory to me that if the facts won't fit in, why so much the worse for the facts is my feeling" (ii, 233). Darwin's close friend, Sir Charles Lyell, after reading the *Origin of Species* suggested that in a future edition, "you may here and there insert an actual case to relieve the vast number of abstract propositions" (ii, 206). It is this absence of "actual cases," of course, which is the entire difficulty with evolution. There are no actual cases.

Chapter VII

FROM BOOM TO RACKET

THE Darwinian gospel on its first appearance in 1859 had an immediate and world-wide success. Huxley in a contribution to Darwin's *Life and Letters* (ii, 179) has written wild nonsense about Darwin earning his place, "by sheer native power, in the teeth of a gale of popular prejudice, and uncheered by a sign of favour or appreciation from the official fountains of honour." The fact of the matter is that the London *Times* on December 26, 1859, devoted no less than the almost unprecedented space of three and a half columns to a highly respectful notice of *The Origin of Species*, this review, incidentally, being written by Huxley himself. The chorus of praise was so general as to drown dissentient voices. Immediately after his book appeared Darwin is found writing to Hooker, "My head will be turned. By Jove, I must try and get a bit modest." The first edition was sold out on the day of issue and new editions and reprints appeared in quick succession.

In 1864 the Royal Society awarded Darwin the Copley Medal, and Dr. Hugh Falconer in seconding the award described the *Origin of Species* as having "instantly fixed the attention of mankind throughout the civilised world" (*More Letters*, i, 255). Within six years of publication of his book Darwin was elected an honorary member of the leading learned societies of eight nations. The position twelve years after the *Origin* appeared was thus described in Professor St. George's Mivart's *Genesis of Species* (p. 10) in 1871 : "It would be difficult to name another living labourer in the field of physical science who has excited an interest so widespread, and given rise to so much praise, gathering around him as he has done a chorus of more or less completely acquiescing disciples, themselves masters of science, and each the representative of a crowd of enthusiastic followers." As noted in an earlier chapter, the last scientist to raise his voice against evolution at the congresses of the British Association for the Advancement of Science was Dr. Wright in 1881. By February, 1888, Herbert Spencer, a rival prophet of evolution, was complaining in the *Nineteenth Century* that "the new biological orthodoxy"—that is, the evolutionist—had become as intolerant as the old, and was rigidly shutting its eyes to everything that did not fit in with the Darwinian idea of natural selection. The booming of evolution remains the most successful scientific stunt on record.

What was the reason for Darwin's success? For one thing, he applied to science the gospel of Malthus which had already become the foundation of political economy. Ricardo's theory of rent and his "iron law of wages," etc., and J. S. Mill's similarly inspired writings all received valuable philosophical reinforcement and expansion from Darwin's theory of natural selection and survival of the fittest in the struggle for existence, with a supposed onwards and upwards march to illimitable improvement and progress.

These ideas of Darwin's also enabled any successful man to justify himself to himself in riding rough-shod over his weaker competitors. Such notions were far more comforting teaching than anything to be found in the sayings of Jesus Christ. As John Morley gracefully expressed it in reviewing the *Descent of Man* in the *Pall Mall Gazette* in 1871 : "Mr. Darwin's work is one of those rare and capital achievements which effect a grave modification in the highest departments in the realm of opinion" (Darwin's *More Letters*, i, 324). Everyone in these highest departments was shown by Darwin to be there by survival of the fittest. The idea was naturally completely acceptable in these elevated quarters. No doubt this was another instance of what Darwin's *More Letters* (i, 71), calls "his supreme power of seeing and thinking what the rest of the world had overlooked." As Darwinism taught that the Bible was all astray in its statements, it also made a wide appeal to those advanced and enlightened people who had had more than enough of the Ten Commandments. There was thus an enormous public ready to receive the comfortable new gospel, and quite willing to overlook the absence of any facts in support of it.

In some quarters there was resistance to Darwin's doctrine. The scientists of France were about the last to be converted, as is recorded in Darwin's *Life and Letters* (iii, 224). It was not until 1878 that Darwin was elected a corresponding member of the French Institute, and then only on the botanical side. He received 26 votes out of a possible 39. In 1872 an unsuccessful attempt had been made to elect him to the zoological side, but he only got 15 votes out of 48. *Nature* of August 1, 1872, quoted an eminent member of the Academy as writing in *Les Mondes*:

"What has closed the doors of the Academy to Mr. Darwin is that the science of those of his books which have made his chief title to fame—the *Origin of Species*, and still more so the *Descent of Man* —is not science, but a mass of assertions and absolutely gratuitous hypotheses, often evidently fallacious. This kind of publication and these theories are a bad example, which a body that respects itself cannot encourage."

In Britain and America there were a small number of scientists who refused to accept evolution. Professor Fleeming Jenkin, an engineer, in an article on the *Origin* in the *North British Review* in 1867 remarked that Darwin's idea that a species could be modified by a favourable variation occurring in an individual was the same as arguing that the arrival of one ship-wrecked European sailor on an island populated by negroes would result in the population gradually turning white in the course of a century or two (vide Mivart's *Genesis of Species*, p. 58). Fleeming Jenkin said he did not anticipate this little difficulty would embarrass the "true believer" in evolution, for : "He can invent trains of ancestors of whose existence there is no evidence; he can call up continents, floods, and peculiar atmospheres; he can dry up oceans, split islands, and parcel out eternity at will; surely with all these advantages he must be a dull fellow if he cannot scheme out a series of animals and circumstances explaining our assumed difficulty quite naturally." (Darwin's *Life and Letters*, iii, 108).

Darwin in the next edition of the *Origin* set to work and tidily patched up this rent in his theory. He said there could be no doubt that owing to similar organisms being similarly acted on by external conditions, "the tendency to vary in the same manner has often been so strong that all the individuals of the same species have been similarly modified without the aid of any form of natural selection." Any theory needed could be supplied on demand apparently. An analysis of the various repairs effected in the six editions of the *Origin* would be instructive.

After reading a well-known work by his fellow evolutionist, Herbert Spencer, Darwin wrote to Hooker in 1866 : "I feel rather mean when I read him; I could hear, and rather enjoy the feeling that he is twice as ingenious and clever as myself, but when I feel he is about a dozen times my superior, even in the master art of wriggling, I feel aggrieved . . ." (*Life and Letters*, iii, 55).

Disbelief in evolution is today regarded as a sign of crankiness and serious mental incapacity. From the five volumes of Darwin's letters one gathers that such disbelief also results in degeneration of character, for those who criticise the great man's views almost invariably appear in these volumes rapidly to develop unpleasant traits in personaility. Sir Richard Owen, the eminent naturalist, commonly suspected of being the author of a severe slating of the *Origin* in the *Edinburgh Review*, turns out to be a most unpleasant person, and "mad with envy because my book has been talked about" (*More Letters*, i, 149). The *Edinburgh Review* article had the audacity to say, "Lasting and fruitful conclusions have, indeed, hitherto been based only on the possession of knowledge; now we are called upon to accept an hypothesis on the plea of want of knowledge. The geological record is so imperfect ! " (*ibid*, i, 146).

Professor St. George Mivart, F.R.S., suffered a similar deterioration of character after acquiring a disbelief in various parts of the
Darwinian theory, expressed by him in review articles and at length
in his book of 1871, the *Genesis of Species*. "He is very unfair," writes
Darwin to Lyell. "You never read such strong letters Mivart wrote to
me about respect to me, begging that I would call on him, etc., etc.
Yet . . . he shows the greatest scorn and animosity towards me, and
with uncommon cleverness says all that is most disagreeable. He makes
me the most arrogant, odious beast that ever lived . . . I suppose that
accursed religious bigotry is at the bottom of it . . ." (*More Letters*
i, 332).

Mivart in his writings had pointed to various gaps in the Darwinian argument. He remarked, for example, on Darwin's objection
to the idea that any special sterility had been imposed on species to
prevent inter-mixture. On this point Darwin wrote in the *Origin:*
"To grant a species the special power of producing hybrids, and then
to stop their further propagation by different degrees of sterility . . .
seems a strange arrangement." Mivart's comment in his *Genesis of
Species* (p. 125) was : "But this only amounts to saying the author
would not have so acted had he been the Creator. 'A strange arrangement' must be admitted anyhow . . . and it is undeniable that the
crossing is checked . . . there is a bar to the intermixture of species,
but not of breeds." This little defect in his theory, as we have seen,
was a standing vexation to Darwin, who objected strongly to Huxley
obtruding it to public notice.

Darwin held natural selection to operate on chance variations
occurring haphazard, and rejected all ideas of design in nature—at
any rate subsequent to the First Cause. Mivart noted Darwin asking,
"Can it be reasonably maintained that the Creator intentionally ordered
. . . that certain fragments of rock should assume certain shapes, so
that the builder might erect his edifice ? " Mivart's comment was
possibly displeasing to Darwin. "It is almost incredible," he wrote,
"but nevertheless it seems necessary to think that the difficulty thus
proposed rests on a sort of notion that amidst the boundless pro-
fusion of nature there is too much for God to superintend; that the
number of objects is too great for an infinite and *omnipresent* Being
to attend to singly to each and all in their due proportions and needs"
(*Genesis of Species*, p. 258). Was this a specimen of Mivart's "accursed
religious bigotry" making Darwin appear as he complained, "the
most arrogant, odious beast that ever lived" ?

In passing it may be noted that one of Darwin's stock arguments
against design in nature was the shape of his nose, which displeased
him every time he looked in the mirror. The point recurs in various
letters. "Will you honestly tell me (and I shall really be much obliged)"
he wrote to Lyell in 1861, "whether you believe the shape of my nose
(eheu !) was ordained and guided by an intelligent cause ? " (*More*

Letters, i, 193). In a letter to Hooker in 1870 Darwin said of the universe, "I can see no evidence of beneficent design, or indeed of design of any kind in the details" (*ibid*, i, 321).

Professor Adam Sedgwick, the geologist, condemned the *Origin of Species* emphatically in the *Spectator* and at the Cambridge Philosophical Society. But the old man was very considerably Darwin's senior, and he was let off by being frozen out of the presence on calling on Darwin some time after his offences. Sedgwick declared quite truly, that the palaeontological record defied the evolutionist at every turn. "I cannot conclude," he wrote in the *Spectator*, "without expressing my detestation of the theory, because of its unflinching materialism; because it has deserted the inductive track, the only track that leads to physical truth; because it utterly repudiates final causes and thereby indicates a demoralised understanding on the part of its advocates. . . Not that I believe that Darwin is an atheist, though I cannot but regard his materialism as atheistical. . . And I think it intensely mischievous. . . Each series of facts is laced together by a series of assumptions and repetition of the one false assumption. You cannot make a good rope out of a string of air bubbles." (Darwin's *Life and Letters*, ii, 298).

Louis Agassiz, the American naturalist, in reviewing the *Origin*, at once put his hand on the weakest spot of all, saying : "Unless Darwin and his followers can succeed in showing that the struggle for life tends to something beyond favouring the existence of some individuals over that of other individuals, they will soon find they have been following a shadow" (*Life and Letters*, ii, 330). Neither Darwin nor anybody else has ever been able to show this, but eighty years have passed without the scientists being able to perceive that they are following a shadow. Agassiz apparently over-rated the penetration of the scientific mind. As to the theory generally, Agassiz said : "Until the facts of Nature are shown to have been mistaken by those who have collected them. . . I shall therefore consider the transmutation theory a scientific mistake, untrue in its facts, unscientific in its methods, and mischievous in its tendency" (*ibid*, ii, 184)—"Have you seen Agassiz's weak metaphysical and theological attack ? " wrote Darwin to Huxley (ii, 330).

The foregoing extracts show that the fundamental defects of the theory of evolution were clearly exposed by competent scientific men at the time of its first appearance. These defects remain today, along with numerous others since disclosed. And yet the public has it rammed down its throat on every hand that evolution is an established scientific fact. In the United States, for instance, when Bryan was campaigning against evolution in 1922, the council of the American Association for the Advancement of Science passed a resolution affirming that evolution was "not a mere guess," and furthermore that "the evidences in favour of the evolution of man are sufficient to convince every scientist of note in the world" (*Nature*, March 3, 1922). This latter statement was not true: that is, unless regarded as announcement of intention by a scientific ring to treat as a blackleg and scab any scientist who rejected evolution. There is evidence from various quarters of evolution today being run as a scientific racket.

Take the case of a leading biologist like Professor Sir William Bateson, who at the Toronto congress of the American Association for the Advancement of Science in 1921 admitted the total failure of all experiments directed to breaking through the natural breeding limits of species. The weight of this negative evidence against evolution could no longer be ignored, he said. While admitting this, Sir William Bateson nevertheless concluded in loyal evolutionist style : "Let us then proclaim in unmistakable language that our faith in evolution is unshaken. Every available line of argument converges on to this inevitable conclusion. . .The difficulties which trouble the professional biologist need not trouble the layman. Our doubts are not as to the reality or truth of evolution, but as to the origin of species, a technical, almost a domestic problem. Any day that mystery may be solved" (*Nature*, April 29, 1922). For his candour in referring in various addresses to the lack of evidence of evolution, Sir William Bateson is stated in his biography written by his son to have suffered at the hands of his scientific brethren a grievous ostracism, amounting to a professional boycott.

Professor G. Macready Price, professor of philosophy and geology at Washington, an anti-evolutionist, is quoted as writing : "When my unorthodox college text-book on geology appeared in 1923 it was pounced upon savagely by the scientific journals—or by all that deigned to notice it at all. I was at that time green enough to think that the editors of these journals would allow me to make some sort of reply to my critics. But I had to learn the sad lesson that my offence had placed me beyond the pale. Finally, some years later, I did manage to get a partial reply to two of my critics in the official journal of the scientific society to which I belonged, but only after I had threatened the editor with the proceedings for libel (*vide Why Be an Ape ?* by a London journalist, Mr. Newman Watts, p. 38).

The way scientists are kicked into line on the subject of evolution was revealed by Mr. Arnold Lunn in his *Flight from Reason* in 1930. He wrote : "The other day I met a Fellow of the Royal Society. 'I am glad you are taking up this issue,' he said, 'because, of course, we professional scientists can do nothing. Our hands are tied. Take my own case for instance. Professor X regards Darwin as a Messiah. He has good jobs in his gift, and no jobs are going excepting to those who worship at the Darwin shrine.' "

Mr. Lunn also quotes the late Professor Thomas Dwight, an eminent anatomist, as saying : "The tyranny of the *zeitgeist* in the matter of evolution is overwhelming to a degree of which outsiders have no idea. Not only does it influence (as I admit it does in my own case) our manners of thinking, but there is an oppression as in the days of the Terror. How very few of the leaders of science dare tell the truth concerning their own state of mind ! "

A palaeontologist who has rejected evolution is Lieut.-Col. L. Merson Davies, a Fellow of the Royal Society of Edinbrugh, a Fellow of the Royal Anthropological Society, and a Fellow of the Geological Society. Colonel Davies in a Victoria Institute paper, said in 1926 : "It 'does not pay,' as they say, to oppose evolution nowadays . . . how many have heard of the words of a leading zoologist like Fleischmann, a scientist of European reputation, who flatly denied that evolution could be regarded as scientifically established ? It is significant that no one has ever undertaken the task of directly opposing Fleischmann; but he was thoroughly abused instead, and soon forgotten. When men of science find that open expression of serious doubts upon the subject is treated after this fashion it is natural that they incline to keep them to themselves. Although the great majority of scientific workers do certainly accept belief in evolution, we have no reason to suppose they all do, even if we seldom hear of them openly opposing it" (vide Dewar's More Difficulties, pp. 124-5).

Mr. Douglas Dewar, F.Z.S., tells how the press has been nobbled up so that the public never hears the truth about evolution. Mr. Dewar was an evolutionist up to 1931, when he published his first book against the theory. In 1912 he was co-author with Mr. F. Finn of an evolutionist book, The Making of Species, which was specially commended by President Theodore Roosevelt. In later life Mr. Dewar whose subject is Indian birds, rejected Darwinism in favour of evolution by sudden mutations. In the end he rejected the entire theory. He writes in his Man, a Special Creation (pp. 103-4) :

"Few people realise how important has been the capturing of the press by evolutionists. Today very few periodicals will publish an article or a paper attacking the evolution theory, and this applies both to the lay and the religious press : most of the religious journals are in the hands of modernists who have accepted the theory of man's animal descent . . . Generally speaking the editors of newspapers believe evolution to be an established fact, and in consequence regard anyone who attacks it as an ignoramus or crank . . . Scientific journals, being conducted by evolutionists, usually decline to publish any contribution that casts even a shadow of doubt on the evolution concept. . . Book publishers . . . are unwilling to publish a book, which as it runs counter to current scientific opinion will either be ignored or savagely attacked. Nor are most of them willing to publish at the author's expense a volume attacking evolution for fear they should lose caste. Thus the public is permitted to hear only one side of the case . . . . the average man . . . is led to believe that evolution is a law of nature as firmly established as is the law of gravity."

As an instance of how Britain's leading scientific journal deals with anti-evolutionists the following facts may be mentioned. The present writer in looking through a file of *Nature* noticed in its issue of November 27, 1937, a half-page review of *Evolution and Its Modern Critics* by Dr. Morley Davies. This book was a rejoinder to Mr. Douglas Dewar's *Difficulties of the Evolution Theory* published in 1931. In 1938 Mr. Dewar replied to Dr. Davies with another book, *More Difficulties of the Evolution Theory*. Inspection of the very full index to *Nature* revealed no review of either of Mr. Dewar's two anti-evolutionist books. Most of *Nature's* review of Dr. Davies' book was occupied with belittlement of Mr. Dewar. Said the reviewer: "In place of a hypothetical discussion, the author takes Mr. Dewar's *Difficulties of the Evolution Theory* in place of a young student's questions." Having thus by inference written Mr. Dewar down to the to the intellectual level of an immature student, *Nature's* reviewer proceeded to give him a kindly lecturette on the extreme rarity of fossilisation, suggesting that if Mr. Dewar watched a dog pulling an animal carcase to pieces he might have some elementary ideas on the subject. In point of actual fact, Mr. Dewar and Mr. Levett-Yeats, as we have noted, had five years before contributed a paper to the Victoria Institute giving detailed statistics on this subject of fossilisation. However, the last thing any reputable scientific journal desires today is to allow anything injurious to evolution to creep into its pages.

Testimony as to a similar state of things in the United States was provided by Mr. Paul Shorey in an article in the *Atlantic Monthly* in 1928. "There is no cause," wrote Mr. Shorey, "that is so immune from criticism, that is so sacred a cow, not only in newspaper offices but in the universities of the North (of America), as Evolution with a capital E. An ambitious young professor may safely assail Christianity or the Constitution of the United States, or George Washington, or female chastity, or marriage, or private property. But he must not apologise for Bryan. . . That would be intolerance, lack of a sense of proportion, failure in open-mindedness, unfaith in progress. It is not done."

In Britain in 1934 an Evolution Protest Movement was formed and presently requested the British Broadcasting Corporation that its president, Sir Ambrose Fleming, F.R.S., might be given an opportunity of stating the case against evolution over the radio. Mr. C. A. Siepmann, the B.B.C. director of talks, refused this, stating in his reply: "It is the policy of the Corporation to allow of reference to evolution in such terms as have the support of the large majority of distinguished scientists in this country." An Evolution Protest Movement leaflet notes that while taking this stand to block an anti-evolution broadcast, the B.B.C. nevertheless defended a communist broadcast on the ground that "there was no greater danger than that a point of view should be suppressed." Curiously enough, it was only by the invention of the thermionic valve by Sir Ambrose Fleming, whom the B.B.C. refused to allow to speak, that radio broadcasting became possible at all. The incident is significant as just one more instance of evolutionist and communistic influences operating in double harness in influential quarters.

Chapter VIII

ALL ABOARD FOR ATHEISM

EVER since its first proclamation eighty years ago, the theory of organic evolution has been actively at work disintegrating the religious beliefs of those who accept it. Its principal achievement has been to empty the churches by mass manufacture of atheists and materialists. Atheism and materialism very frequently find their political embodiment in communism. It is not correct to say that all evolutionists are atheists, materialists, and communists. They are simply headed that way, that is all. It is correct on the other hand, to say that communists are almost invariably atheists, materialists, and evolutionists. Evolution is essential to materialism and atheism in that it provides a mechanical explanation of the universe without any spiritual principle.

Both Darwin, prophet of evolution, and Huxley, his high priest, had abandoned belief in Christianity at the time they took up with evolution. Darwin records that after his return from the voyage in the Beagle in 1836, at which date he was twenty-seven years of age, "I gradually came to disbelieve in Christianity as a divine revelation " (*Life and Letters*, i, 308). Huxley similarly relates that by 1850, at the age of twenty-five, he had "long done with Pentateuchal cosmogony," and desired "some particle of evidence" that animals came into being by creation (*ibid* ii, 187-90). That neither held established religion in especial esteem is evident by their letters. Darwin in 1859, for instance, wrote to Hooker complaining that a certain reviewer of his book "drags in immortality, and sets the priests at me," and is ready to "tell the black beasts how to catch me" (*ibid*, ii, 228). Huxley's attitude was well known, and right at the end within a year of his death he wrote in 1894 : "I am not afraid of the priests in the long run. Scientific method is the white ant which will slowly but surely destroy their fortifications," and lead to "the gradual emancipation of the ignorant upper and lower classes, the former of whom especially are the strength of priests." (Huxley's *Life and Letters*, ii, 379). The word "priests" in these extracts is used as descriptive of clergy generally, irrespective of denomination.

Huxley expounded his theological views publicly and emphatically in his addresses and books. Darwin did not do so. In various letters he is to be found stating, "My theory is in a muddle." The fact of the matter seems to be that his own intuitive feelings for a long time prevented him from carrying his scientific views to their logical conclusion. According to the Darwinian theory of natural selection all living things are the product of chance variations without purpose or design. Modern evolutionists carry the idea a stage further, regarding life itself as a chance product of inorganic matter. Darwin for a long time refused to accept this idea, but in 1871 he is to be found imagining a chance generation of life from non-living matter—"in some little warm pond, with all sorts of ammonia and phosphoric salts, light, heat, electricity, etc." (*Life and Letters*, iii, 18).

As he advanced in life, Darwin became more completely materialistic in his views. Less than two months before his death he wrote on February 28, 1882 : "Though no evidence worth anything has as yet, in my opinion, been advanced in favour of a living being, being developed fròm inorganic matter, yet I cannot avoid believing the possibility of this will be proved some day in accordance with the law of continuity. . . Whether the existence of a conscious God can be proved from the existence of the so-called laws of nature (*i.e.*, fixed sequences of events) is a perplexing subject, on which I have thought, but cannot see my way clearly" (*More Letters*, ii, 171). The chaotic nature of Darwin's views is revealed in the following statement made in 1879 :

"The old argument from design in *Nature*, as given by Paley, which formerly seemed to me so conclusive, fails now that the law of natural selection has been discovered. . . There seems to be no more design in the variability of organic beings, and in the action of natural selection than in the course which the wind blows. . . At the present day the most usual argument for the existence of an intelligent God is drawn from the deep inward convictions and feelings which are experienced by most persons. . . Formerly I was led by such feelings ...to the firm conviction of the existence of God and the immortality of the soul. . . I well remember my conviction that there is more in man than the mere breath of the body. But now the grandest scenes would not cause any such conviction and feelings to rise in my mind. Another source of conviction in the existence of God. . . follows from the extreme difficulty, or rather impossibility, of conceiving this immense and wonderful universe, including man and his capacity for looking backwards and far into futurity, as the result of blind chance or necessity. . . This conclusion . . . has gradually, with many fluctuations, become weaker. . . I, for one, must be content to remain an Agnostic" (*Life and Letters*, i, 309-12).

In 1881 Darwin was asked if a certain correspondent was correct in describing him as saying everything was due to chance. Darwin replied that the word "chance" must have been used "in relation only to purpose in the origination of species." He added : "On the other hand, if we consider the whole universe the mind refuses to look at it as the outcome of chance—that is, without design or purpose. The whole question seems to me insoluble, for I cannot put much or any faith in the so-called intuitions of the human mind, which have been developed, as I cannot doubt, from such a mind as animals possess; and what would their convictions or intuitions be worth?" (*More Letters*, i, 395).

Darwin's position was that whatever the First Cause of all things might be, everything thereafter occurred by chance. Alfred Russell Wallace and Sir Charles Lyell accepted evolution subject to in-breathings of creative power to help the process on, particularly with respect to the appearance of man upon the earth. Wallace regarded natural selection as quite incapable of accounting for man's powers of speech, his taste for music, mathematical faculty, and other attributes separating him by an immense gulf from the lower animals, and in his *Darwinism* in 1901 Wallace imagined evolution helped along by creative power from time to time just as the direction of a curve is imperceptibly altered. Lyell in correspondence with Darwin in 1859 had raised the same point of injections of primeval creative power, and Darwin replied, "If I were convinced that I required such additions to the theory of natural selection, I would reject it as rubbish" (*Life and Letters*, ii, 210).

Huxley, with his endless capacity for writing nonsense, rebuked the critics of Darwin for "the most singular" of their fallacies, "that which charges Mr. Darwin with having attempted to re-instate the old pagan goddess, Chance" (Darwin's *Life and Letters*, ii, 199). Nevertheless, one finds Sir Edward Poulton in his presidential address to the British Association in 1937 recalling the "heroic help" given in the study of Darwinian natural selection by Miss Welldon "who four times recorded the result of 4,096 throws of dice, showing that the faces with more than three points were, on the average, uppermost slightly more often than was to be expected" (*Nature*, Sept. 4, 1937). According to evolutionists, chance developed the human eye by the operation of discontinuous light on a freckle on the skin. Chance is the only God evolution knows.

Huxley, chief propagandist of Darwinism, expressed himself with frequency and vigour on theological matters. He saw not a tittle of evidence, he said, of a Deity standing to mankind in the relationship of a father. "I am unable," he declared, "to discover any 'moral' purpose, or anything but a stream of purpose towards the consummation of the cosmic process, chiefly by means of the struggle for existence, which is no more righteous or unrighteous than any other mechanism" (Huxley's *Life and Letters*, i, 241, and ii, 303).

Darwin's moral ideas were based on his evolutionist doctrine. He expressed the opinion in 1879 that "most or all sentient beings have been developed in such a manner, through natural selection, that pleasurable sensations serve as their habitual guides" (*Life and Letters*, i, 310). Morality on this basis slides easily and naturally into the doctrine of Aleister Crowley and the Black Mass, 'Do as thou wilt, there is no other law.' The so-called new morality runs along these lines, and the same idea (wrapped in cotton wool) is at the back of the new education, etc.

While Darwin became increasingly materialistic in his views in later life, Huxley became less so. However, one finds Huxley writing on the subject of morality as follows in 1892 : "So far as mankind has acquired the conviction that the observance of certain rules of conduct is essential to the maintenance of social existence it may be proper to say that 'Providence,' operating through man, has generated morality" (Huxley's *Life and Letters*, ii, 303). Huxley in his Romanes lecture of 1893 admitted that the ethical principle of disregard of self is opposed to the self-regarding principle on which evolution is assumed to have taken place. Huxley also admitted that natural selection fails to account for man's taste for music and his innate sense of moral beauty.

Enough has been quoted of the views of Darwin and Huxley to show the deteriorating effect of their evolutionist imaginings on their entire outlook on life. They are adrift from whatever bearings they ever had, and without any idea of where they are. Their moral standards, so far as they have any clearly defined standards, are inferior to the Christian ones they threw away. Their idea of man as descended from brute beasts is a degraded conception as compared with the conception of man as a spiritual being; and the Darwinian idea of the living world as a chance product without aim or purpose is one which runs counter to the intuitions of mankind. If what they had to tell us was true we should have to accept it and reconcile ourselves to it as best we could. But the standards of proof accepted by evolutionist scientists would be despised by a mediaeval witch finder. The late Lord Halsbury, a former Lord Chancellor of England, has been quoted as very truly saying : "In court we are expected to give full proof in support of every assertion. A professor, on the other hand, appears to consider himself relieved of any such anxiety" (*vide Why be an Ape?*, p. 34).

Evolution is an emanation of darkness, a product of a decadent age. A passage in Huxley's *Life* reveals the progress of that decadence. In expounding his theological views in correspondence with Charles Kingsley in 1863, Huxley remarked that, except with two or three scientific colleagues, he felt himself helplessly at variance with his fellowmen on these matters, and as remote from them as they would feel among a lot of Hottentots. He added : "I don't like this state of things for myself—least of all do I see how it will work out for my children" (Huxley's *Life and Letters*, i, 240). Huxley need have had no fear. The evolution boom swept him and his family onward and upwards. Nine years after he wrote the University of St. Andrews recognised his eminence by electing him its rector. Oxford University later acclaimed him as one of the great men of the day by bestowing an honorary D.C.L. Finally, in 1892 Queen Victoria (or her Ministers) thought it not inappropriate for the Defender of the Faith to summon this active and noisy propagator of disbelief to membership of her Privy Council.

As for the Huxley family, they have flourished mightily in preaching the same Huxleian gospel. In 1899 the evolutionists, in their task of emancipating the ignorant upper and lower classes from thraldom to religion, founded the Rationalist Press Association. Seven years later that body was busy pushing out in scores of thousands a book of 432 pages sold at the obviously unprofitable price of sixpence, and proudly flaunting on its cover an excerpt from the *Sunday School Chronicle* quoting the Bishop of London as saying that this particular book "has done more damage to Christianity during the last few years than all the rest of the sceptical books put together." Inside the cover appeared the name of Professor Leonard Huxley, son of the great Huxley, as one of the executive officers of the association.

In the third generation Huxley's grandson, Professor Julian Huxley, is able triumphantly to proclaim in his widely circulated *Essays of a Biologist* (p, 74) that "any view of God as a personal being is becoming frankly untenable." The "march of knowledge," he affirms, has left no room in the universe for any such idea. "Creation of earth and stars, plants, animals and man—Darwin swept the last vestiges of that into the wastepaper basket of outworn imaginations, already piled high with the debris of earlier ages," so affirms this worthy scion of the House of Huxley.

It is worth noting that evolution is the mainstay of the Rationalist Press Association volume of which Professor Julian Huxley's father was one of the sponsors. The volume, *The Churches and Modern Thought* by Philip Vivian, solemnly assures us without a quiver of an eyelid that "Evolution is no longer a mere speculative theory, possibly or probably true, but an established fact accepted by the whole scientific world with hardly a single dissentient voice " (p. 169). In putting Christianity through the mincing machine, the author dwells on the pitiful lack of evidence on the religious side. "Even the working man," he tells us, "will not remain satisfied with a theology which maintains the necessity for a foundation of facts, and yet is unable to prove them" (p. 361). With gusto he quotes on page 339 the following statement from T. H. Huxley's essay on "Science and Morals" : "The foundation of morality is to have done, once and for all, with lying; to give up pretending to believe that for which there is no evidence. . ."

If evolutionists were to follow Grandfather Huxley's advice, and give up pretending to believe that for which there is no evidence, not very much would be left of their theory. Even prominent evolutionists in pronouncements from the throne in the scientists' parliament have declared evolution to be just as much a matter of faith as religion. The eminent Professor D. H. Scott, for example, in his presidential address to the botanical section of the British Association in 1921 is to be found saying : "Is then evolution a scientifically ascertained fact ? No ! We must hold it as an act of faith because there is no alternative." Eight years later Professor D. M. S. Watson, of the University of London, told the assembled zoologists of the British Association in 1929 : "The theory of evolution is a theory universally accepted, not because it can be proved true, but because the only alternative, special creation, is clearly incredible."* The facts assembled in these pages suggest strongly that Professors Scott and Watson had excellent grounds for saying no evidence can be found of evolution. The fact of the matter is simply that the Huxleys, Vivians, and the rest, approach evolution in a spirit of credulity and religion in a spirit of incredulity.

It is noticeable that both Professor Julian Huxley and Mr. Vivian attach importance to the number of believers as an item of proof of evolution. Professor Huxley, as previously noted, opens the 1929 *Encyclopaedia Britannica* article on evolution by telling us that "among competent biologists and geologists, there is not a single one who is not convinced," etc. Mr. Vivian proclaims evolution as "accepted by the whole scientific world with hardly a dissentient voice." On the other hand, we have noted various qualified scientific men stating that it 'does not pay' today to express disbelief in evolution, and that the scientific billets only go to professing evolutionsts.

The proposition thus seems to run : (1) Evolution is established fact. (2) It is established because scientists believe in it. (3) Scientists have to believe in evolution to rise on the payroll. The reader can draw his own conclusion from these premises. There is certainly no pecuniary sacrifice attaching to belief in evolution today. It is a suitable belief for a scientist desirous of presently possessing a higher-grade motor car. And pursuit of truth for truth's sake is a cold, cheerless occupation compared with going places in a limousine. You travel in pursuit of truth on a secondhand push-bicycle.

"Materialism," says Professor John Elof Boodin, in his *Cosmic Evolution* (1925), "has substituted magic for sober thought. The whole process of evolution becomes a succession of miracles without intelligible ground in the process. The appearance in a world of chance of any order at all, the emergence of life with its series of forms and organs, the final appearance of intelligence and a sense of beauty— all are miracles. . . That any age should take seriously such an incoherent mixture of mysticism and science is evidence of nothing so much as a want of logical thinking. . . By some magic the antecedent

---

*Professors Scott and Watson are quoted in Major Wren's *Evolution: Fact or Fiction?* pp. 22 and 59. *Nature's* reports of their addresses, apparently condensed, are to similar effect in less direct language.

forms are supposed to yield new forms and characters. By chance variations the structure of protoplasm is supposed to be built up from inorganic matter, and by further chance the various forms and characters appear. . . Chance is God. . . Materialism offers the most astounding instance of credulity in history" (extracts quoted in Major Wren's *Evolution, Fact or Fiction ?*).

Evolution is not a science at all. It is a religion, and a very low-grade religion, with its hymns played in jazz and syncopated cacophony, and its sanctuaries adorned with cubist art—the religion of the Godless, of the crazy intellectuals who don't know anything about anything except knowledge. A few faint gleams of light begin to appear through the murk of this evolutionist hell-upon-earth. A stray scientist here and there begins to perceive that knowledge is not all. Professor Sir Arthur Eddington, peering out into space through his astronomical telescope, perceives something else looming into view. "The problem of knowledge," he writes, "is an outer shell beneath which lies another philosophical problem—the problem of values" (*Philosophy of Physical Science*, 1939, p. 222).

With their heads stuffed full of what they call knowledge our materialist-evolutionist intellectuals have lost all sense of values. They forget that the human rational intellect is but a measuring machine capable of reporting on the difference between this and that, but dealing all the time with nothing but the symbols of reality. Not all the books and all the professors in the world can even tell you the taste of bacon and eggs, you have to eat a dish of them to find that out. Rationalism gives you the menu card and calls it a dinner. It hands you the music score and calls it a concert. Your contact with reality comes through other parts of your consciousness.

Rationalism cannot measure God Almighty, cannot weigh Him : therefore God Almighty does not exist. But as William James, the psychologist, points out* the innumerable men and women who in all ages have felt the presence of God behind phenomena need no logical demonstration to prove that God exists : they *know* that all the words in all the creeds and religions of the world are but the feeblest shadow of a tremendous fact. Not all men have this insight, but the intuitive feeling that there is an omnipresent power in the universe, outside and beyond man, yet friendly to him and his ideals, has endured in the human heart through all ages and in all lands.

The evolutionist-materialist says these intuitions are superstitious delusions. He wants logical proof before he believes in anything but his material world. But the materialist has yet to be born who can provide any logical proof that the material universe has any existence apart from his own consciousness of it. When it comes to logical foundations he is as hopelessly bankrupt as his opponent. It all comes down to the practical working value in human life of the goods that

---

* *Varieties of Religious Experience*, pp. 73 and 443.

evolutionist-materialism and religion have to offer. We hear a lot about "Democracy" today. A very valuable book entitled *Modern Democracies* was written a few years after the war of 1914-1918 by the late Lord Bryce. After surveying the democracies of the world as they then stood, Lord Bryce expressed the opinion that the future of civilisation depends upon the future of religion. Of Christianity he said :

"Christianity—a far more powerful force than any political ideas or political institutions, since it works on the inmost heart of man—has produced nearly all the moral progress that has been achieved since it first appeared, and can in individual cases transmute lead into gold; yet Christianity has not done these things for peoples, because checked or perverted by the worse propensities of human nature, it has never been applied in practice" (vol. ii, p. 585).

What has Darwinian monkey-man materialism produced since it first appeared ? Nothing but dirt and degradation all along the line. No one is twopence the better for it. What is it based upon? Upon nothing whatever but faith, upon belief in the reality of the unseen—belief in the fossils that cannot be produced, belief in the embryological evidence that does not exist, belief in the breeding experiments that refuse to come off. It is faith unjustified by works—"fanatic faith, that wedded fast to some dear falsehood, hugs it to the last." Whether we seek our ancestors in the Garden of Eden or the monkey-house, it is faith that guides us equally to the one quarter or the other : upwards to the stars and immortal harmonies beyond this muddy vesture of decay; or downwards to the dark earth and the beasts that perish.

Chapter IX

EVOLUTION'S OFFSPRING

IN CURRENT thought today everything is assumed to be in a state of evolution into something else. The theory of organic evolution is not only treated as a fact, but has been made the basis of a philosophy which has invaded every branch of knowledge, and every side of life. This assumed evolutionary change is associated with the idea of progress and an inevitable onwards and upwards movement. The net result of the propagation of evolutionary ideas is, first and foremost, to lower the public resistance to change. The propagation of evolution is thus of the highest advantage to those who, for their own reasons —which are not necessarily those publicly proclaimed by them—wish to effect a revolutionary overthrow of existing institutions. Their battle is half won if they can succeed in propagating the idea that everything is in a state of inevitable and automatically beneficial evolution. In modern parlance, evolution is first-class revolutionary sales talk.

The pioneer in this evolutionary philosophy—himself very far from being a revolutionary—was Herbert Spencer. He embarked on the undertaking a few years before Darwin's *Origin of Species* appeared, and devoted most of a long life to it. The results were embodied in the ten ponderous volumes of his *System of Synthetic Philosophy*, treating of first principles, biology, psychology, sociology, and ethics. It is significant that in applying evolutionary principles to the universe at large, Herbert Spencer passed over inorganic nature. He admitted that logically a treatise on evolution in inorganic nature should have followed next after his *First Principles*, but excused himself on the ground that "even without it the scheme is too extensive," and also evolution in organic nature is "of more immediate importance" (*Epitome of Synthetic Philosophy*, p. 65).

An additional reason for the omission of this portion of the Spencerian philosophy may possibly have been the fact that science appears to be unable to discover any trace whatever of the assumed universal, onwards and upwards, integrative principle of evolution at work in inorganic nature. "Whatever a star's evolution may be," says Sir James Jeans in *The Universe Around Us* (p. 306), "it is essential that it should always be down the steps : any upward step is impossible." The sun, for example, is reported by science to be always losing energy without full recovery. Turning from stars to atoms, the scientific report is that the largest and heaviest among the atoms (such as radium and uranium) are continually breaking up of themselves; and some atoms have been broken up artificially. But nothing is known of any process of building up the higher-class atoms out of simpler elements. Dr. F. Soddy, professor of physical chemistry at Oxford University, in his contribution to *Evolution in the Light of Modern Knowledge* (pp. 401-4) says there is no reason to suppose the simpler atoms came first and the complex ones afterwards, or that one atom is the parent of another. "Can one honestly see even a trace of that consecutive progress. . . reflected in the ways of the material universe ? " he asks.

In discussing this matter of inorganic nature and evolution in his book *The Bible Confirmed by Science* (from which the above excerpts are taken), Dr. W. Bell Dawson noted the contention in 1930 of Dr. R. A. Millikan that "cosmic radiation" indicates that atoms are being built up in inter-stellar space. Dr. Dawson remarked that such a view could not be definitely verified, and that Sir James Jeans in his *Mysterious Universe* (pp. 75-6) showed the probablilities to be enormously against it. That the Millikan idea remains purely supposititious is indicated from *Nature's* report of Professor P. M. S. Blackett's Kelvin Lecture on "Cosmic Rays" before the Institution of Electrical Engineers on April 27, 1939. Professor Blackett is reported as saying, "There is little exact knowledge about these rays, and it is assumed that they are due to radiation from the extra-solar universe." If it is only assumption that the cosmic rays come from the extra-solar universe, it presumably requires an extra large dose of evolutionist faith to spring the further assumption that cosmic rays are evolving atoms there. Except for this unverified idea, it seems to be correct to say that science knows nothing of evolution in inorganic nature.

It thus appears that Herbert Spencer might have included a treatise on "The Principles of Inorganic Evolution" in his vast work with no more labour than was involved in the famous chapter Ixxii "Concerning Snakes," in Horrebow's *Natural History of Iceland*, the whole truth of which reads exactly thus : "There are no snakes to be met with throughout the whole island." The reader will recall that the great Darwin himself, after reading a volume of the Spencerian philosophy described Herbert Spencer as "about a dozen times my superior, even in the master art of wriggling." Nevertheless, if evolution is a universal principle it should be able to explain how material things came to be. Professor Sir William Bateson admitted this, saying "Every theory of evolution must be such as to accord with the facts of physics and chemistry : a primary necessity to which our predecessors paid small heed" (*vide* Dr. W. Bell Dawon's *Is Evolution True* ? No. 2, p. 8). Once again our evolutionist steed collapses at the hurdle.

If the great integrative principle of evolution could not be applied to inorganic nature, it was, however, applied to everything else. The evolutionist dram-drinker soon found himself "seeing things," and with respect to man in particular. Huxley, accepting evolution as a fact, and accepting man's descent from the lower animals as a fact, found himself in sight of various illuminating consequential facts. As dispenser-in-chief of the Darwinian distillation, the aged Huxley in 1893 solemnly got on to his feet, and rammed the following statements down the throats of the callow undergraduates of Oxford in his Romanes Lecture of that year :

"Man, the animal, in fact, has worked his way to the headship of the sentient world, and has become the superb animal he is, in virtue of his success in the struggle for existence . . . the self-assertion, the unscrupulous seizing upon all that can be grasped, the tenacious holding of all that can be kept, which constitute the essence of the struggle for existence, have answered. For his successful progress, throughout the savage state, man has been largely indebted to those qualities which he shares with the ape and the tiger; his exceptional physical organisation; his cunning, his sociability, his curiosity and his imitativeness; his ruthless and ferocious destructiveness when his anger is aroused by opposition . . . after the manner of successful persons, civilised man would gladly kick down the ladder by which he has climbed."

There are no "ifs" and "ands" in this Huxleian pronouncement. All is as real and visible as the pink snakes crawling over the counter-pane. Unfortunately, the archaeologists, looking in the ground for their facts instead of drawing upon their imaginations for them, appear to be finding evidence that civilised man was in existence all the time. One finds an eminent archaeologist, Dr. A. H. Sayce of the University of Oxford, quoted as saying : "Neither in Egypt nor in Babylon has any beginning of civilisation been found. As far back as archaeology can take us, man is already civilised, building cities and temples, carving stone into artistic form, and even employing a system of picture writing; and of Egypt it may be said, the older the country, the more perfect it is found to be. The fact is a very remarkable one, in view of the modern theories of development, and of the evolution of civilisation out of barbarism. Whatever may be the reason, such theories are not borne out by the discoveries of archaeology. Instead of the progress we should expect, we find retrogression and decay; and where we look for the rude beginnings of art, we find an advanced society and artistic perfection. Is it possible that the Biblical view is right after all, and that civilised man has been civilised from the out-set ? " (vide Dr. W. Bell Dawson's The Bible Confirmed by Science, p. 141).

In Germany, the modernists in the Church and the socialists outside it, both took up with Darwinism at an early date. On it also Frederick Nietzsche built up his philosophy of Prussianism. "Nietzsche was the child of Darwin," says Dr. Will Durant in his Story of

Philosophy. Darwin, Durant says, swept away the last remnant of the theological basis of modern morals, and "men who could think clearly soon perceived . . . that in this battle we call life, what we need is not goodness but strength, not humility but pride." Nietzsche contended that just as man had evolved from an ape-like ancestor, so a higher type "superman" would evolve from man. Here are a few Nietzschean pearls of wisdom :

"We . . . believe that [man's] will to Life had to be intensified into unconditional Will to Power; we hold that hardness, violence, slavery, danger in the street and in the heart, secrecy, stoicism, arts of temptation and devilry of all kinds; that everything evil, terrible, tyrannical, wild-beast-like and serpent-like in man contributes to the elevation of the species just as much as the opposite—and in saying this we do not even say enough."—*Beyond Good and Evil,* sec. 44.

"Ye say it is the good cause which halloweth even war ? I say unto you, it is the good war, which halloweth every cause."—*Thus spake Zarathusa,* "War and Warriors."

"One must . . . resist all sentimental weakness : life is in its essence appropriation, injury, the overpowering of whatever is foreign to us and weaker than ourselves, suppression, hardness, the forcing upon others of our own forms, the incorporation of others, or, at the very least and mildest, their exploitation."—*Beyond Good and Evil,* sec. 259.

"We children of the future . . . do not by any means think it desirable that the kingdom of righteousness and peace should be established on the earth . . . we count ourselves among the conquerors; we ponder over the need of a new order of things, even of a new slavery—for every strengthening and elevation of the type 'man' also involves a new form of slavery."—*The Joyous Wisdom,* sec. 377.

Huxley held cunning and ruthlessness to be the means by which man had ascended from beast, according to the Darwinian hypothesis. Nietzsche arrived at exactly the same idea from the same source, and as a practical-minded person urged his fellow Germans to carry on with the good work and joyously evolve into the "Super-man," "the blond beast, lustfully roving in search of booty and victory." Darwin's and Huxley's countrymen have since been engaged in two expensive efforts to remove this Darwinian inspiration from the German intellect.

Marxian socialism, like its stable companion atheism, with which it is usually to be found jogging along in double harness, has drawn sustenance freely from the evolutionary nosebag. Karl Marx (original family name Mordecai) produced his original programme of revolutionary violence in his *Communist Manifesto* of 1848, and the first volume of his theoretical justification of it, *Das Kapital,** in 1867. In this he drew largely on Malthusian ideas, via Ricardo. His financial

---

*It was to Darwin that Marx wished to dedicate *Das Kapital.* With characteristic tact, Darwin declined the honour. (Editor).

backer and collaborator, Engels, in his celebrated oration over the grave of Marx, declared Marx and Darwin to be twin discoverers of the law of evolution, saying: "Just as Darwin discovered the law of evolution of organic nature, so Marx discovered the evolutionary law of human history—the simple fact that . . . the production of the material necessities of life and the corresponding stage of economic evolution of a people or a period provides a foundation upon which the national institutions, legal systems, art, and even the religious ideals of the people in question have been built, and upon which, therefore, their explanation must be based."

Numerous quotations showing the close tie-up between evolution and revolutionary socialism appear in Mr. Dan Gilbert's excellent little book *Evolution: the Root of All Isms* from which the foregoing is quoted. Enrico Ferri, the Italian socialist leader, asserted in his *Socialism and Modern Science* that Marxian socialism "is nothing but a vital and logical corollary in part of Darwinian, in part of Spencerian, evolution." Said Karl Kautsky, German socialist leader: "For Marx, on the other hand, the class struggle was but a particular instance of the universal law of evolution, whose essential qualities are in no case peaceful." The American socialist, Morris Hillquit (real name Hilkowicz) said in his *Socialism in Theory and Practice:* "Karl Marx alone consistently introduced the spirit of Darwinism into the study of social phenomena by substituting the . . . doctrine of the class struggle in the more modern stages of social development for the . . . doctrine of the struggle for existence in its lower stages." Hillquit proceeds to lay it down that "in the ascending scale of organic existence the struggle between individuals of the same species gradually abates and is superseded by collective struggles of such individuals."

The various socialist writers, after imbibing at the evolutionary fountain, arrive at very similar views to those of Nietzsche as to the morality of the whole business. Survival is the evolutionary test of fitness, and anything which helps to survival is moral. Here are a few flowers of socialist thought along these lines of evolutionist morality:

"All the factors which impede the path to its [socialism's] approximate realisation are anti-ethical or immoral; contrariwise, all factors or movements which tend in its direction are ethical."—Morris Hillquit, noted American Socialist, in his *Socialism in Theory and Practice.*

"The socialist has a distinct aim in view. If he can carry the initial stages towards it by means of the count-of-heads majority, by all means let him do so. If, on the other hand, he sees the possibility of carrying a salient portion of his programme by trampling on this majority, by all means let him do this also."—E. Belfort Bax, noted English socialist, in his *Ethics of Socialism.*

"The dictatorship of the proletariat is nothing else than power based upon force and limited by nothing—by no kind of law and by absolutely no rule."—Lenin, *Complete Works*, vol. xviii, p. 361.

"The class-war is not only to continue, but must be carried on with considerably increased intensity. The classes inimical to the workers are, it is true, beaten, but the individuals continue to exist. They have not all of them been shot as yet, not all of them have been caught, not all of them have been physically annihilated. The carrying out of this task lies before us."—Krylenko, principal public prosecutor of the Soviet Union, in May, 1934.

The foregoing excerpts appear among others in Mr. Gilbert's book to which reference has been made. They could be multiplied indefinitely. Everything is based upon the evolutionist line of thought of the survival of the fittest in the struggle for existence—that which succeeds, anything that pays, is moral. The socialistic authors quoted exhibit exactly the same moral ideas as underlie Nietzsche's Prussianistic philosophy, and both are derived from the same source.

The tie-up between evolutionary scientists and Bolshevism is notorious. According to London press statements of July, 1940, Professor J. B. S. Haldane, noted evolutionist, was then chairman of the editorial board of the *Daily Worker*, official organ of the Communist Party of Great Britain, and which, according to Lord Newton in the House of Lords on June 13, 1940, had "done all it can to impede the war effort of the Government."

In his references to the file of Britain's leading scientific journal *Nature*, the present writer noticed in its issue of May 23, 1936, three and a half of its foremost pages devoted to a glowing review of a new book, *Soviet Communism** by Sidney and Beatrice Webb. This book is a whitewashing of Bolshevism by two English socialists who have spent the last half-century in percolating socialism, via the Fabian Society, into the British upper classes, civil service, and universities. The book was outside the usual range of topics of a purely scientific journal, and the present writer in turning over the files of *Nature* did not chance to encounter reviews of books critical of Bolshevism. *Nature* declared the Webbs' book to be "a work of ultimate significance to the English-speaking world"; and the Bolshevik revolution itself is described as "an event in human history like the spiritual emancipation that came with Christianity and the political emancipation that culminated in the French Revolution." This scientific journal thus showed itself ready to go out of its way to use its pages

---

*Beatrice Webb in "Our Partnership," p. 491, categorically affirmed that *Soviet Communism* was the work of her husband and herself. But the one-time Soviet Army officer, Colonel Bogolepov, who fled to the West, giving evidence before the Internal Security Sub-Committee of the United States Senate Committee of the Judiciary on April 7th, 1952, stated that the entire text of the Webbs' book had been prepared in the Soviet Foreign Office. The Colonel explained that he had done most of the ghost-writing himself in the line of duty.
(Editor).

as a vehicle for communist propaganda in the university and other circles in which it circulates.

Evolutionist worship of Bolshevism is reciprocated by Bolshevist worship of evolution. Evolution is a mainstay of the Soviet anti-God campaign. One Soviet evolutionist effort for the "spiritual emancipation" of Russia was recorded in Mr. R. O. G. Urch's *Rabbit King of Russia* (pp. 82-83). He relates that according to the *Krasnaya Gazette* of December, 1925, Professor Ilya Ivanovich had left with a scientific expedition on a mission to the Congo. The object was to catch a number of female chimpanzees, and "to fertilise the apes by artificial methods and bring back the mothers with the little human apes to gladden the hearts of the Anti-God Society in Soviet Russia and prove that "There is no God." The result remained obscure, a detailed rumour in circulation in Moscow the next year stating that the ship on which the expedition was returning had been lost with all hands. Evolutionist experiment obstinately refuses to come off, not even to help along the spiritual emancipation of Soviet citizens.

Among the varied offspring of the Darwinian doctrine of man's animal descent are the numerous "scientific" doctrines of free-love. Mr. Gilbert deals at length with these in his useful *Evolution: the Root of all Isms*. He points out, for example, that the late Professor Freud in his *General Introduction to Psychoanalysis*, of which teaching he was the founder, lays it down that "man's animal nature is ineradicable" and makes this the justification for "giving full rein to one's sexuality." This Freud affirms to be quite in order as evolution shows that man is "an animal accustomed to the freedom of the jungle," and "unadapted" to the restrictions of Christian civilisation. This is the entire burden of the psychoanalytic gospel. A popular American university textbook, *An Outline of Psychoanalysis*, edited by J. S. Van Teslaar, states : "Psychoanalysis represents but an extension of the theory of evolution, an application of the principle of evolution . . ." Dr. Samuel D. Schmalhausen, probably America's most popular and persistent champion of the so-called new morality of "sex expression" and "sex experimentation" with unlimited license, bases his arguments throughout on animal behaviour and man's supposed animal descent. "The sexual revolution is the terminal phase of the scientific revolution," he says in his *Sex and Civilisation*. Robert Briffault, another leading writer on these lines, in a contribution to a symposium by America's "sex philosophers," entitled *Our Neurotic Age*, records various observations of the habits of monkeys, etc., and triumphantly announces : "No indication is to be found among any animal of an appreciation of the rudiments of Christian morals." Christian morals thus lack evolutionary sanction, and are defined by Briffault as a "disease" of civilisation.

Another by-product of evolution dealt with by Gilbert is Determinism, which teaches that it is wrong to put criminals in gaol because human beings are not responsible for their actions and have no freewill. Metchnikoff, leading Determinist, in his *Nature of Man* says : "Evolution knows nothing of free will. All our actions, scientifically speaking, are the necessary outcome of chemical processes of the brain." A number of evolutionists take this stand. Behaviourism goes a step further again, and flatly denies the existence of the soul, of free will, and of consciousness. Its founder, Dr. John B. Watson, started out with a study of animal psychology and then carried on with human psychology. Behaviourism maintains that man is entirely animal, and that "man should be treated just as the animal is treated."

In concluding this brief survey of some of the fruits of the theory of organic evolution, it is interesting to recall the comment of Adam Sedgwick, the geologist, on the Darwinian theory when it first appeared. After reading through his copy of the *Origin of Species*, the old man sat down and wrote a letter to Darwin on Christmas Eve. 1859. In it he said :

"We all admit development as a fact of history : but how came it about ? Here in language, and still more in logic, we are point-blank at issue. There is a moral or metaphysical part of nature as well as a physical. A man who denies this is deep in the mire of folly. 'Tis the crown and glory of organic science that it does through *final cause*, link material and moral . . . You have ignored this link; and if I mistake not your meaning, you have done your best in one or two pregnant cases to break it. Were it possible (which, thank God, it is not) to break it, humanity, in my mind, would suffer a damage that might brutalise it, and sink the human race into a lower state of degradation than any into which it has fallen since its written records tell us of its history" (Darwin's *Life and Letters*, ii, 248).

For great masses of people the theory of organic evolution has undoubtedly broken the link between material and moral, has had brutalising results, and has plunged them into a lower state of degradation. It represents lost ground in every direction.

Chapter X

WHAT SCIENCE IS FINDING

NOTHING is more striking in recent scientific literature than the steady accumulation by evolutionists themselves of a mass of facts wholly opposed to the theory of evolution to which they continue to affirm their adherence. This applies not only in the field of biology, but also, as will be seen in the next chapter, in archaeology and anthropology. In every direction the scientists themselves are busy removing bricks and undermining the walls of the citadel of evolution while at the same time loudly proclaiming the fabric to be intact. The indications are either that the entire theory will presently be abandoned as obsolete and untenable, or, alternatively, that science will be reduced to the teaching of dogma regardless of fact, in which direction it has already proceeded a considerable distance.

Evolution teaches the mutability of species. Modern scientific observation is becoming increasingly impressed with the great stability of species as the outstanding fact about living organisms. Darwin in concluding the *Origin of Species* proclaimed that there was no essential difference between varieties and species, species being all the time in process of evolution into new forms, and varieties being merely incipient new species. "We shall at least be freed from the vain search for the undiscovered and undiscoverable essence of the term species," he declared. Living organisms being nothing but a series of dissolving views with everything in transformation, Darwin laid it down that in future naturalists in labelling species would only have to decide whether any form happened to be "sufficiently important to deserve a specific name."

This idea is in line with a notion freely advanced in intellectual quarters today : that there are no pure races of mankind, and that it therefore follows there is no difference between a white man and a negro. On these lines we had Mr. Bernard Shaw holding forth in South Africa a few years back that the uplift of that country would be brought about most quickly by the two million whites intermarrying with the seven million black inhabitants of the country.

On the Darwinian view individuals alone exist in nature, species being purely an arbitrary concept. Nevertheless, species seemed to be sufficiently real to Darwin for him to devote a book of 700 pages to trying to account for their origin. Louis Agassiz, in reviewing

Darwin's book in the *American Journal of Science* for July, 1860, commented on this anomaly, and asked : "If species do not exist at all, as the supporters of the transmutation theory maintain, how can they vary ? and if individuals alone exist, how can the differences which may be observed among them prove the variability of species ? " It does not appear that Agassiz ever got an answer to this conundrum.

According to evolutionist teaching, living things are a chance product of inorganic matter, and are being continuously pruned and moulded into new shapes by their environment. The living forms are, so to speak, mere jelly or plastic putty, struggling together for existence, in a given environment, and by natural selection attaining an endless succession of forms increasingly adapted to the environment. The environment is the determining factor, and heredity amounts to little or nothing. Such is the grand theory. When the scientists lay their theoretical spectacles aside we find them announcing the exact opposite as the outstanding fact to be observed in the natural world.

Here, for example, is a recent report from the scientific throne, delivered in 1939 by the president of the zoological section of the British Association, Professor James Ritchie, of the University of Edinburgh : "We are accustomed to lay stress upon the variation of living things, upon which evolution depends, but surely more remarkable is the stability of living organisms, which retain their own characters in spite of changes in the environment, and whose germ cells pass these characters unaltered through countless generations" (*Nature*, September 25, 1939).

As far back as 1921 Professor D. H. Scott was saying exactly the same thing as president of the botanical section of the British Association for that year. He said : "The small variations on which the natural selectionist relied so much, have proved, for the most part, to be merely fluctuations oscillating about a mean, and therefore incapable of giving rise to permanent new types. . . The whole tendency of modern work is to show that in living things heredity is supreme. An organism is what it is by virtue of the constitution of the germ plasm derived from its parents" (*Nature*, September 29, 1921).

In the Oxford University *Evolution* book of 1938, one finds Mr. J. Z. Young, demonstrator in zoology at Oxford, saying (p. 180) : "As Henderson puts it, 'living things preserve or tend to preserve an ideal form, while through them flows a steady stream of energy and matter which is ever changing, yet moulded by life; organised in short.' The very object of our study as biologists is this organisation and its preservation, it is 'the first fact which strikes us about organisms,' and it is, as Woodger remarks, curious that it should be necessary to draw attention to it."—This is tacit admission that evolutionist theory has blinded scientists to what stares them in the face.

Similar testimony to this constancy of species is borne by Professor A. W. Lindsey, of Denison University, U.S.A., in his interesting *Problems of Evolution* published by Macmillans in 1931. Professor Lindsey remarks that : "All of the selection theories, all ideas of isolation, all of our knowledge of mutations, serve only to show that the characters which make up a species may be reassorted, re-distributed, preserved in part and destroyed in part, or modified to some degree" (p. 134).—In other words, you can get variations within species, and sometimes wide variations, but you cannot get anything whatever which was not there to begin with.

All the above statements are made by professing evolutionists. Nevertheless, all are testimony unfavourable to the theory of evolution. Professor Ritchie after remarking on the stability of species, gets no further than observing that "we must conceive of evolution as a process of extreme slowness." Failure to discover evidence of evolution has led to the scientists making very large drafts on the bank of time. On this point Sir William Bateson remarked a good many years ago that "Time cannot complete that which has not begun."

The majority of the evolutionists did not share this Batesonian view. They felt that evolution could achieve anything if supplied with sufficient time. A furious battle raged in the scientific world from 1892 to 1921 over this subject of time. The geologists, calculating the rate of deposition of the sedimentary rocks, had in the main been willing to supply generous allowances of time. Sir Archibald Geikie on this basis provided evolution with about 100 million years. The physicists next upset the applecart. Lord Kelvin asserted that not more than 40 million years had elapsed since the molten earth solidified. Professor Tait knocked this down to 10 million. Professor Joly eased the situation by estimating the sodium content of the oceans and the amount of salt carried in by the rivers, and he calculated the age of the oceans as from 80 to 90 million years. Professor Sollas in a British Association presidential address reprinted in his *Age of the Earth* (1908) affirmed that, properly regarded, both the sedimentary rocks and the oceans testified to about 26 million years as the correct figure : and he asserted that this should be sufficient time for everybody. Most of these estimates were revised by those who made them, sometimes out of recognition.

At the British Association meeting in 1905, Sir George Darwin threw out a life-line to the evolutionists by announcing that Madame Curie's investigations with radium might throw light on the age of the earth. Calculations were thereupon embarked on as to the rate of disintegration of radio-active ores in the rocks, and by 1921 Lord Rayleigh had evolution nicely fixed up with about 1000 million years available for something to evolve into something else. A few more hundred million years have since been thrown in to prevent any undue cramping. Professor Ritchie in his 1939 address was able to remark that "now a concensus of opinion admits credibility to estimates based upon the break-up of radio-active minerals in the rocks," and that they might say life had existed on the earth for perhaps 1200 million years, and that the birth of the sun and stars took place about 2000 million years ago.

Mr. Dewar in his *More Difficulties* (pp. 101-6) remarks that having abandoned belief in evolution, he has ceased to be particularly interested in the time allowance for the imagined process. He points out that the present fashionable calculation based on radio-activity is in total conflict with all the previous calculations, and seems to have been jumped at on the principle of not looking a gift horse too closely in the mouth. He notes Professor Joly, Mr. D. J. Whitney, and Professor A. Holmes recording the most completely discordant results from tests of radio-active ores obtained from the same rocks. He remarks that the present theory assumes that the rate of disintegration of radio-active ores has always been the same as today; whereas Lord Rutherford, Professor Joly, and Professor Fermi have expressed the opinion that the radio-active elements may be simply the end-product of other elements which disintegrated so rapidly that no trace of them now remains on earth.

Sir Ambrose Fleming, F.R.S., in a presidential address to the Victoria Institute in 1935, said, "It is certain that the geologists have not found any generally agreed and unquestionable test by which to determine the geological age in years of earth crust materials or deposits and the assumptions made by some are disputed by others." Sir Ambrose Fleming has since continued to insist emphatically on the worthlessness of present evolutionist chronology, and he appears to have good grounds for doing so. In every sphere it touches evolution demands the wholesale manufacture of conclusions to fit in with preconceived theory.

Although now supplied with years by the thousand million, the main result seems to be that science finds species remaining unchanged over longer periods than was the case under its former chronology. This stability continues to be the great fact presented both to the zoologist and the botanist. Coupled with it is also another impressive fact, the endless small variations within the species themselves. Species exhibit all manner of varieties and strains within the interbreeding community, and within these varieties and strains again are endless small individual differences. Here in Nelson, New Zealand, we had the eminent Dutch botanist and evolutionist, Dr. J. P. Lotsy tell us in his Cawthron Lecture in 1927 that : "It is practically impossible to make a group of identical individuals; we can but make a group of similar individuals, because in nature no two individuals are alike in all respects." Many other naturalists have made similar remark.

"To my mind," says Mr. Dewar, "one of the most impressive phenomena of the organic world is this variety, coupled with stability of the type. It would seem that in a sense every individual is a new creation" (*More Difficulties*, p. 72). Mr. Dewar adds that he knows of no better explanation of this variety than that given by Paley away in 1802 in his *Natural Theology* (p. 170), when he wrote that what we see around us might "induce us to believe that *VARIETY* itself, distinct from every other reason, was a motive in the mind of the Creator, or with the agents of His will." In Paley's day not so much was known about this variety in individuals as is now the case. Nobody, for example, had then recognised that apparently the whorls on the finger tips are not alike in any two human beings.

So far from evolution having relieved naturalists of the task of deciding what is a true species and what is not, one finds them still busy discussing the point. In 1938 the Linnæan Society of London celebrated its 150th anniversary with a symposium on "The Concept of Species from the time of Linnaeus to the Present Day." From the London *Times* report (May 26, 1938) it appears that the speakers were agreed that species are a reality, and that a number of them were of the opinion that the true concept of species is an interbreeding community. This, of course, is sliding back a long way from Darwin's idea that in the light of evolution a species is merely a group of similar individuals of sufficient importance to deserve a specific name—a purely arbitrary conception, in short.

As to what are true species—using the word as connoting interbreeding communities—it appears that science today has very limited knowledge. In the Linnean Society discussion Dr. Karl Jordan of the Tring Museum pointed out that a large number of described species are known only from a specimen or two, and great numbers of them are probably mere varieties within other species. In the Oxford University *Evoluion* book (p. 108) Dr. O. W. Richards remarks that there is only one way to obtain genetical information about species, and that is by breeding experiments. He points out, however, that "many animals are extremely difficult to breed in captivity, and in any case there is no hope that more than a few of the very numerous known species of animals can be investigated genetically."

To a layman it appears curious that with the scientists all at sixes and sevens as to what constitutes a species, they should all be so dogmatic on the *origin* of species. The cart seems to have got before the horse. Mr. Dewar as an anti-evolutionist and a believer in special creation adopts a more rational attitude. In his *More Difficulties* (p. 10) he says he does not in the least profess to know what the units of creation are. He does not assert that every species, every genus, or even every family has been specially created. He holds it the duty of biologists to try to discover what these units are. In his opinion the data at present are not nearly adequate to make even a tentative pronouncement; and "all that can be safely said is that so far, breeding experiments seem to indicate that the units of creation are small, or in other words, that the number of these units is great."

It is interesting to note that Professor St. George Mivart in his

Genesis of Species in 1871 quoted various passages from the writings of the early Christian Fathers which showed them as holding it allowable to believe that existing forms of animals and plants are not necessarily the forms of their original creation, but are derived therefrom. Mr. Dewar from his anti-evolutionist point of view says he sees no harm in scientists adopting evolution as alternative hypothesis to creation, but, he adds, "I consider it suicidal to adopt evolution as a creed, to distort facts to cause them to conform to it, to brush aside facts not amenable to it, and to minimise difficulties" (*More Difficulties*, p. 205).

The task of science is to trace back the chain of causation as far as she can, preceding step by step from fact to fact. The theory of organic evolution is assertion unsupported by evidence, and is an invasion by science of the domain of philosophy and religion. In tracing back causation science must in any case come to a point beyond which she cannot go. Man's rational intellect has its limitations, and is quite incapable of conceiving a First Cause. There are only three possible ideas as to the origin of life : (1) that it was created by external agency, (2) that it has always existed, and (3) that it is the product of spontaneous generation. Modern evolutionist science rejects the idea of creation as "incredible." Major Wren in his *Evolution, Fact or Fiction?* quotes various pronouncements on the point. Says Professor J. B. S. Haldane, "The evidence for the existence of a superhuman Designer . . . was invalidated by the discovery of evolution and the theory of Natural Selection" (An Address to *Modern Churchmen*, Oxford, 1926). Sir Arthur Keith is quoted as stating, "To say that God made matter, and out of dead matter made living matter, cannot satisfy even a child's intelligence, for the child's next question is sure to be, 'and who made God?'"

In point of actual fact the constancy of species—which we have noted as impressing itself so strongly on the scientific mind of late—coupled with the endless variety in the individual organisms, is much more consistent with the idea of special creation than with evolution. The observed fact that apparently no two individuals in nature are exactly alike in all respects is in harmony with the intuition of mankind that every living thing is a direct and unique manifestation of the creative power of an omnipotent and omnipresent Deity. When the evolutionist rejects the idea of special creation as "incredible" the most he can say against it is very much what Mr. Dewar in his *Man: a Special Creation* (p. 95) quotes Dr. S. Zuckerman as saying with respect to human beings :

"Either evolutionary change or miraculous divine intervention lies at the back of human intelligence. The second of these possibilities does not lend itself to scientific examination. It may be the correct explanation, but, from the scientific point of view, it cannot be legitimately resorted to in answer to the problem of man's dominantly successful behaviour until all possibilities of more objective explanation through morphological, physiological and psychological observation and experiment are exhausted" (*Functional Activities of Man, Monkeys and Apes*, p. 155).

The scientists are not content to adopt a neutral attitude. They are not content to state the plain truth that science has nothing to report about the origin of species, and that the quest for evidence of evolution has run to a dead end with negative results in every direction. They insist on ramming evolution down the public throat, evidence or no evidence. In doing this they cease to be scientists and become theologians. The majority of present-day scientists are atheistically inclined, and evolution provides them with a philosophical background for their atheism. The basic issue is thus not scientific but theological.

When the history of modern evolutionist theory is studied its atheistical basis stands out conspicuously throughout. Any reference book in outlining the history of the theory will be found making reference to Buffon (1707-1788), Erasmus Darwin (1731-1802), Lamarck (1744-1829), and finally Charles Darwin (1809-1882). The Comte de Buffon was a prominent figure among the French philosophers and men of letters inveighing against established religion and providing the ideas which were presently put into practical application in the French Revolution. Guizot in his *History of France* describes Buffon as "absolutely unshackled by any religious prejudices"; and notes him pulling strings to avoid having his *Histoire Naturelle* black-listed by the ecclesiastical authorities. An old encyclopaedia (Chambers, 1885) describes him as "largely participating in the vices of the time," and his widow was the last of the numerous mistresses of the Duke of Orleans before that prince was guillotined in the revolution. Buffon put forward evolutionist views, and declared life and mind to be a property of matter.

Lamarck's religious views are not mentioned in any reference book at hand, but he was under Buffon's patronage and was tutor to Buffon's son for a number of years, and presumably found the atmosphere of the household to his taste. Alison's *History of Europe* (i, 176-7) states : "Almost the whole of the philosophical and literary writers in Paris, for a quarter of a century before the French Revolution broke out, were avowed infidels; the grand object of all their efforts was to load religion with obloquy, or, what was more efficacious in France, to turn it into ridicule. When David Hume was invited at Paris to meet a party of eighteen of the most celebrated men in the French capital, he found, to his astonishment, that he was the *least* sceptical of the party : he was the only one present who admitted even the probability of the existence of a Supreme Being." It does not appear whether Buffon was of this company, but he ornamented many such gatherings.

Erasmus Darwin, grandfather of Charles Darwin, aired evolutionary views in his *Zoonomia* published in 1794. He was a physician at Derby and became acquainted with Rousseau during the period when the latter was living in exile in England at Lichfield under the patronage of David Hume, and corresponded with him thereafter. Rousseau was the chief philosopher of the French Revolution, affirming that man was born innocent and good, that the savage was the model of every virtue, and that all miseries and vices were due to the tyranny of kings, the deceptions of priests, the oppressions of nobles, and the evils of civilisation. Property, he declared, was the great evil which had ruined mankind : and self-control was a violation of nature. It was significant of much that was to come that Rousseau opened his famous *Discourse on the Origin of Inequality* by saying, "Let us begin by laying facts aside as they do not affect the question."

Charles Darwin in his *Life and Letters* (iii, 179) describes his father, son of Erasmus, as "a freethinker in religious matters," and although he himself was at one time a divinity student at Cambridge, he presently turned his back on both the Church and Christianity. It is noticeable that both Darwin and Huxley abandoned their religious beliefs at or about the time at which they adopted evolutionist views.

Darwin derived the basic idea of his theory—the struggle for existence—from Malthus, and it is curious that although Malthus himself was a clergyman of the Church of England, his father, Daniel Malthus, according to Beale's *Racial Decay* (p. 38) was a friend and executor of Rousseau. If Buckle's statement in his *History of Civilisation* is correct, and Voltaire was the real originator of the Malthusian idea of population increasing faster than food supply, we have yet another root of modern biological science running back into the midst of French pre-revolutionary philosophy. "Voltaire," says Guizot's *History of France*, "has remained the true representative of the mocking and stone-flinging phase of free-thinking . . . At the outcome of the bloody slough of the French Revolution and from the chaos it caused in men's souls, it was the infidelity of Voltaire which remained at the bottom of the scepticism and moral disorder of the France of our day" (Sampson Low's edition, 1889, p. 521).

The foregoing facts are sufficient evidence of a pronounced atheistical background in the incubation of the theory of organic evolution. It is obvious that if belief in God is rejected, a necessity at once arises for some theory accounting for the origin of life otherwise than by creation from the dust of the earth. Evolution meets this need, and the indications throughout are that evolution is a theological and not a scientific product. It was invented to meet the requirements of atheism, and it is maintained and propagated for the same reason.

Chapter XI

MAN AND CIVILISATION

THAT expensive compendium of evolutionary fairy stories, the fourteenth edition of the *Enclyclopaedia Britannica*, opens its article on civilisation with the announcement that "there could be no real understanding of the fundamental characteristics of civilisation until the fact was well established and digested that if we could trace back man's lineage far enough we should find it merging into that of wild animals." The truth of the matter is that modern research into the origins of civilisation and human culture is busy all the time piling up a mass of facts totally inconsistent with any such evolutionist ideas.

The archaeologists excavating in Babylonia and Egypt report that they are unable to discover any beginnings to civilisation. As far back as they can go man is already civilised. At the same time the diffusionist school of anthropologists which has grown up in the past twenty years reports that the weight of evidence is that no savage race ever invented anything, and that all culture was diffused from a common source in South-western Asia, with its centre somewhere about the head of the Persian Gulf.

On page 71 above we noted the late Professor A. H. Sayce remarking on this failure of archaeology to discover any beginnings to civilisation, and asking whether the truth was that the Bible was right after all, and man was civilised from the beginning. In quoting this passage in his book Dr. Bell Dawson did not mention its source, but it is possibly from the Huxley Lecture of 1930 on "The Antiquity of Civilised Man" which Professor Sayce delivered to the Royal Anthropological Society, and of which an abridged report appears in *Nature* of November 29, 1930. In that report one finds Professor Sayce remarking incidentally that the jewellery discovered in the remains of the earliest civilisation would grace a Bond Street jeweller's window today. When the present writer, in common with the many other New Zealand soldiers of the last war, visited the Cairo Museum, nothing impressed him more than the beautiful workmanship of the jewellery of the First Dynasty, dating back to a period almost as far before Moses as we are after him.

Writing of the earliest remains of pre-dynastic civilisation in Egypt, Mr. Arthur Weigall, formerly inspector-general of antiquities to the Egyptian Government, says in his *Short History of Ancient Egypt* (1934, p. 19) : ". . . We find ourselves confronted with a civilisation in being and we really do not know whence it came. Writing had begun; the arts were already highly developed; great armies were in commission; cities had grown up; and the king was surrounded by his ministers and his nobles." Other archaeologists make similar report as to the absence of any discoverable primitive beginnings to these ancient civilisations.

Turning to the side of anthropology, we find a recent writer of the diffusionist school, Lord Raglan, who was president of the anthropological section of the British Association in 1933, providing considerable further food for thought. In his *How Came Civilisation?* (1939, pp. 56-7) he states that all the evidence is that the inventions and discoveries on which European civilisation—that is to say, Graeco-Roman civilisation—was based, seem to have been made somewhere within a region centering on Persia and extending from Egypt to North India and China. "The question then arises," adds Lord Raglan, "were the people of Persia, Mesopotamia, etc., when they began to make all these discoveries and invent all these traits, savages ? The answer must be that if they were they must have been very different from any savages, either ancient or modern, of whom we know anything, since these latter, as we must repeat, are not known ever to have invented or discovered anything."

Current belief today is that the savage races of the world are in the same state civilised man is supposed to have been in a few thousand years ago, and if left alone would ultimately rise to civilised status by process of gradual "evolution." Lord Raglan in summing up the diffusionist case in his most interesting little book, says all the evidence is in exactly the opposite direction, and that "no savage society, when left to itself, has ever made the slightest progress." The only change that has ever been observed to take place in these isolated societies is a change for the worse.

Many lands are stated by Lord Raglan to provide evidence of this retrogression, and it is especially evident in Polynesia. Language and customs show that the people of the Pacific Islands from Hawaii to Easter Island and across to New Zealand probably spread from some common centre. They possess sea-going canoes, but since they have been known to Europeans they have never ventured far out of sight of land. Their ancestors must have made long voyages again and again. In many of the islands are erections built of large blocks of stone. The modern Polynesian is completely ignorant of the art of building in stone. Dixon in his *Building of Cultures* (p. 280) says that "the Polynesians in their eastward drift into the Pacific lost textiles, pottery, and metal-working, and gave up the use of the bow." Fragments of pottery are found scattered about on islands where the natives have now lost the art. In the New Hebrides the natives had the art of weaving in the seventeenth century, but have since lost it. Lord

Raglan devotes a chapter to instances of deterioration of culture among different savage races, and says that so far as he can learn there is nothing whatever to put into the scale against it. Savage races are capable of being civilised by missionary effort, but are incapable of civilising themselves.

Taking such items as the bow and arrow, the domestication of animals, the plough and the hoe, pottery and the potter's wheel, etc., Lord Raglan contends that all the evidence points to diffusion of these inventions and discoveries from a common centre. "I know of no case," he says, "in which anything which can be described as an invention has been recorded as having been made by a living savage . . . People who themselves have never had an idea exhibiting the slightest sign of originality have no difficulty in crediting primitives or savages with brains of the utmost fertility. . . what have Binks the banker and Brown the bus-driver invented? There are in our midst thousands of intelligent and capable Binkses and Browns who have invented nothing whatever; can it really be believed that every savage community, however small and primitive, has produced a succession of men possessing an inventive genius such as has been totally denied to Binks and Brown? " (pp. 40-1).

Throughout historic times we know that civilisation has spread by diffusion. On what ground, asks Lord Raglan, are we to assume that in pre-historic times the exact opposite was the case and everything was independently invented? Everything points to things like the baking and glazing of pottery, the use of the potter's wheel, etc., etc., only being discovered once. The distribution of culture points to groups of people pushing out in all directions from the original home of mankind. The larger the cultural equipment in such a movement, the further it would be likely to get, and adaptation to the environment would probably consist of dropping whatever elements were unsuitable to life in the wilderness. On top of this, as families possessing skill in various crafts became extinct further cultural loss would be probable. In another striking passage Lord Raglan says :

"We know that our own civilisation in all but its latest phases, was not evolved locally, but derived from the Mediterranean. We know that Greece derived its civilisation from Asia Minor, Crete and Egypt. We, like the Greeks and Romans, have improved upon the civilisation which we received from outside, but it is quite untrue to say we evolved our own civilisation. It is then clearly not the fact that civilisation has *everywhere* been evolved out of savagery, and to say that it has *anywhere* been evolved out of savagery is a guess which cannot be supported by any evidence. As Niebuhr (quoted by Tylor, *Primitive Culture*, i. 41) remarked, 'no single example can be brought forward of an actually savage people having independently become civilised.' So far as we know, all civilisation has been derived from pre-existing civilisation, not from savagery.

"Of the real beginnings of culture we know nothing for certain, and it is very doubtful whether we ever shall. It seems likely that the cradle-land of the human race was in South-western Asia, where was also the seat of the earliest civilisations, yet there are fewer traces of 'primitive man' there than in many other parts of the world. Whether this is because the earliest cultures are beneath the silt of the Euphrates or the Indus, or whether their remains still await the chance disturbance of the surface at some hitherto unsuspected spot we cannot say. What we can say is that all the facts alleged as the beginnings of culture are fallacious" (pp. 50-1).

It is impossible to present here even an outline of the evidence on which the diffusionist anthropologists reach their conclusions. Lord Raglan's intensely interesting little book so freely drawn upon above, gives the most recent outline of the case; and the reader desiring more information will find it in the books of Dr. W. J. Perry and the late Sir Grafton Elliot Smith, formerly professor of anatomy in the University of London, who appear to have been the leading exponents of this line of research. It may be mentioned in passing that the diffusionist contention is that the Maya and Inca civilisation of America was carried from East Asia by sea-voyagers across the Pacific. Like the ancient civilisations of the Near East, this appears to have no primitive beginnings; and the diffusionsts point to many remarkable affinities with Asiatic civilisation.

Coupled with the inability of the archaeologists to uncover any primitive origins of the Mesopotamian and Egyptian civilisations, the evidence assembled by the diffusionist anthropologists presents believers in human evolution with some very considerable nuts to crack. Curiously enough, the diffusionists themselves are all convinced evolutionists. The late Professor Elliot Smith ranked as a leading authority on monkey-men. Lord Raglan in his book says on page 56 that "we may suppose that man was evolved from the ape within this region" of ancient culture. Dr. Perry in his *Growth of Civilisation* (1924) on page 112 reveals himself as an evolutionist also. His view is that through a misunderstanding of evolutionary doctrine it has been assumed that simple forms of social organism must necessarily have preceded the more advanced in all parts of the world. Both he and Professor Elliot Smith claim in their books that Egypt provides evidence of a development of civilisation from primitive beginnings. The archaeologists, however, appear unable to find evidence either there or anywhere else of the origins of any of the ancient civilisations.

Accepting evolution in general, the diffusionists reject all idea of evolution as an operating principle in human society. We have seen the flimsy and worthless evidence which is supposed to show mankind as descended from brute beasts. On top of this the archaeologists are unable to find any trace of civilised man ascending from barbarism. Finally, to crown all, the diffusionist anthropologists present a mass of facts all pointing to the diffusion of culture throughout the world from some common centre. They may overstate their evidence in some respects, but they certainly assemble together enough to make an exceedingly strong *prima facie* case against the idea of any upward evolutionary movement from barbarism to civilisation. The net result is complete demolition of any idea whatsoever of evolution in mankind and human affairs. In view of all the facts the following from Mr. Dewar's *Man: a Special Creation* (p. 28) is to the point :

"We have no reason for supposing that . . . the mental powers of the pre-historic men known to us were lower than those of their descendants; in these early races of man, to quote Professor W. Schmidt (*European Civilisation*, 1934, p. 76) : 'No "ape-like" features are to be found. On the contrary, their really human character manifests itself with purity, clearness and beauty, as certainly as anywhere else in the whole history of mankind . . . Thus, once and for all, we may finally abandon any expectation of fresh evolutionary links being established between the spiritual life of man and that of the highest forms of animal life. Even in the earliest representatives of mankind known to us, the soul is so absolutely and completely human that the advance to it from the highest level of the brute creation is more plainly than ever seen to be an impossibility, and mental development such as evolution requires it utterly excluded.' "

Lord Raglan points out that except for the complex known as Western civilisation, all the cultures of the world today are in a state of decay and degeneration. Civilisation, he says in concluding his book, is far from being a process which keeps going on everywhere. It is really an event which has only occurred twice on the grand scale. All the evidence suggests that the first time was somewhere in Southwestern Asia about the fourth millenium B.C., at which time a number of discoveries were made—corn-growing, cattle-breeding, metalworking, pottery, the wheel, the sail, the loom, the brick—which discoveries were diffused in varying degree about the world. This civilisation reached its prime, and then stagnated and decayed, finally collapsing with the fall of the Roman Empire, the Romans themselves inventing practically nothing.

After a thousand-year interval a new burst of enterprise and inventiveness came, and our present Western civilisation began to arise. "Our Western civilisation," says Lord Raglan, "is not a product of evolution or any other natural process, but the result of a series of historical coincidences" (p. 181). In 1400 Europe knew very little that had not been known in Babylon, Egypt, Greece and Rome. In the fifteenth century four very important events occurred. The Turks took Constantinople, and the dispersion of the scholars of the Byzantine Empire flooded Europe with classical knowledge; America was discovered and a new world opened up; block printing was introduced from China; and a ferment of religious thought accompanied the rise of Protestantism. The upshot of these happenings was a stirring among mankind which has continued to our day, with steam, electricity, and all manner of inventions transforming life throughout the world. How the first civilisation arose we do not know; but Lord Raglan points out that it was religious in its inspiration, whereas the second was mainly secular. Other writers, however, dwell on the fact that learning was preserved through the Dark Ages in the monasteries and by survival to the fifteenth century of the Christian Byzantine Empire; that the synods of the Church provided the pattern on which the representative political institutions of Europe were modelled; and that the Christian tradition gave Europe a totally different civilisation from the despotisms of Asia, all kingly power being viewed under it as held in stewardship from God.

The diffusionist anthropologists, as already remarked, are evolutionists who flatly reject evolution as a factor in human affairs. Evolution apparently operates throughout the rest of nature, turns apes into men, and then suddenly ceases to operate for reasons not explained. The diffusionist literature is unsatisfactory in that fact and speculation are as badly mixed up as in the rest of present-day scientific literature, and it is not always easy for the lay reader to discover where one ends and the other begins. The diffusionists are orthodox scientifically in their views in so far that evolution is evolution up to man's appearance, and religion is apparently superstition wherever appearing. Nevertheless, the actual facts they present as to diffusion of culture from a common centre are numerous, striking, and destructive of belief in human evolution.

The late Sir Grafton Elliot Smith in his *Human History* (1934 edition, pp. 59-60) is even more emphatic than Lord Raglan in his rejection of evolution as a factor in human affairs. The idea that human destiny is under control of the terrestrial forces of nature he regards as a mistaken and fallacious application of science to history. He protests strongly against the almost mystical significance given by some writers to climate and geographical environment as implying "some inevitable process of mechanically working development leading to inevitable results in shaping human qualities and behaviour." History, he says, is cataclysmic, and he quotes with high approval an article to this effect by Sir Charles Oman, former professor of modern history at Oxford, in the *National Review* for February, 1929.

In the history department at Oxford University the student under Professor Oman apparently had to leave his evolutionist ideas outside on the doormat for the time being. Professor Oman is quoted as saying: "Two generations have now passed since the blessed word 'Evolution' was invented, and was applied as a universal panacea for all the problems of the universe—historical no less than physical or metaphysical. By this I mean that a whole school of historians have set forth the thesis that history is a continuous logical process, a series of inevitable results following on well-marshalled tables of causes." Sir Charles Oman will have none of this. The career of mankind, he says, has been shaped by accidents and catastrophes and by the action of dominating personalities who have deliberately provoked great movements, peaceful and warlike, which have shaped the destiny of the world.

"The great events in Human History," says Professor Elliot Smith, "were provoked by individual human beings exercising their wills to change the direction of human thought and action, or by natural catastrophes forcing men of insight to embark on new enterprises." He points out that, according to Sir Charles Oman, America would have been colonised by the Norse inhabitants of Greenland if the Black Death in 1350 had not completely destroyed these people. Eventually, the irruption of the Turks into Europe blocked the old trade routes to Asia, and led to search for a new route to the Indies. These historical incidents provided the predisposing causes for a momentous event. But it was the vision and persistence of Columbus which effected the transformation.

The importance of the individual is similarly stressed by Lord Raglan in his *How Came Civilisation?* (p. 172). "Sentimentalists," he says, "may imagine that new culture forms arise from the 'communal mind' or the 'spirit of the folk' or some such abstraction, but the fact is that new ideas can only occur to individuals and do only occur to highly exceptional individuals." Lord Raglan incidentally remarks that it is often assumed that decay is due to conservatism. But decay can occur in other ways than by standing still or going backward. "It is often less realised on the other hand," he adds (p. 172), "that decay of culture can be brought about even more rapidly by breaking away from the past; by the belief that we could and should go back to nature, shaking off the burden of tradition and all that it entails, and living and developing in the innocent freedom of primitive man. People who think like this fail to realise that man became man by getting away from nature, and that it is unnatural not merely to cook food and wear clothes, but to read and write, and even to speak. We learn these arts not from nature, but from tradition. The belief that primitive man was wiser and better than we are is really a symptom of degeneration, of 'that degeneration of the democratic theory which imagines that there is a peculiar inspiration in the opinions of the ignorant' (John Buchan, *Augustus*, p. 340)."

An outstanding feature of the diffusionist argument is the emphasis placed on individual enterprise and initiative as the source and mainspring of civilisation. Equally outstanding is the disregard of the individual in the evolutionist interpretation of history, of which interpretation the Marxian socialist materialist view of history is an offshoot. The dominant idea of evolutionist philosophy as applied to sociology and history is the insignificance of the individual as compared with society past and present. In his *Critical Examination of Socialism* (1909, chap. viii) W. H. Mallock discussed this point at length. He remarks that the modern socialists did not originate this leading idea of theirs, but borrowed it from the evolutionists, among whom Herbert Spencer was its most systematic exponent.

Herbert Spencer in his *Study of Sociology* (p. 35) lays it down that the great man is only the "proximate originator" of changes, and that the real explanation of these changes must be sought in the "aggregate of conditions" in which he exists. Mallock points out that Macaulay in his essay on Dryden said the same thing : " it is the age that makes the man, not the man that makes the age." This idea was the mainstay also of Buckle's *History of Civilisation*. Edward Bellamy in his *Looking Backwards*, a description of a socialist utopia which had an enormous sale half a century back, likewise asserted that "nine hundred and ninety-nine parts of the thousand of the produce of every man are the result of his social inheritance and environment." Herbert Spencer in his book quoted goes at length into demonstrating that the inventor of a new printing press just installed by the London *Times* was no more than its "proximate initiator," and that the press was really produced by the "aggregate conditions" of the period. This aggregate of conditions similarly produced Shakespeare's plays.

Benjamin Kidd, a semi-socialist, is noted by Mallock as helping along the good work by pointing out that various inventions have been arrived at almost simultaneously, and "thus rival and independent claims have been made for the discovery of the differential calculus, the invention of the steam-engine, the methods of spectrum analysis, the telephone, the telegraph, as well as many other discoveries." It is thus inferred that almost anybody might make these discoveries. Mr. Mallock remarks that actually the position with many inventions is that a number of men are trying to scale a peak at the same time, and it is not surprising that two or three men of exceptional ability should sometimes simultaneously reach the previously virgin summit. That anybody might have made the invention is no more demonstrated by this happening than an ascent of the Matterhorn demonstrates that all the people in the tourist hotel at the bottom could have made it. Yet this is the burden all the time of the evolutionist-socialist song—the individual is nothing, environment is everything.

Mankind is viewed in the mass in these theories. But when anything practical is needed in life it is the individual man that counts. The patient at death's door is not helped by being told that man is a great physician and having the first passer-by taken in to attend to him. A series of great frescoes to adorn a public building is not secured by information from the sociological department that great artists are the product of the aggregate of conditions : millions of men subject to this aggregate of conditions might be taken off the streets and tried in turn, and nothing result but hopeless daubs.

The point to which the evolutionist-socialist argument is directed is to show the smallness of the products which the able man can really claim as his own. Another point that it seeks to make is the commonness of ability, which is regarded as purely a product of the environment. Furthermore, whether the great man is a rare or a common phenomenon, his inventions and discoveries become common property. Mallock observed Mr. Sidney Webb (now Lord Passfield, the pope of the Fabian socialists) giving a practical turn to the argument. He noted Mr. Webb, discussing the question of equal pay for all, and holding that the able man has no moral right to a greater share of the product of labour than the less able worker. If one man's brains and effort are responsible for nine-tenths of the value produced, and the other man's one-tenth, they are thus each entitled to fifty per cent. Mr. Webb is quoted as saying that this proposal "has an abstract justification, as the special energy and ability with which some persons are born is an unearned increment due to the effect of the struggle for existence upon their ancestors, and consequently, having been produced by society, is as much due to society as the unearned increment of rent."

Such is evolutionist-socialist philosophy applied to economic affairs. Mr. Webb's notion is not the aberration of a person of no account. It is the idea of one of the most influential socialists in the British Empire. The same idea forms the entire theme of a big book written by another eminent socialist, Mr. G. Bernard Shaw. In his *Intelligent Woman's Guide to Socialism and Capitalism* (1928), Mr. Shaw says equality of income is the basis of socialism, and on page 341 he says, "when there is a difference between the business ability of one person and another, the price of that difference is rent," and this "rent" must be "nationalised" by equal pay for all. Duffers and loafers under socialism are thus entitled to more than they produce : able and energetic workers, on the other hand, have no moral claim to the larger or superior produce of their labour, which is an unearned increment due to accidental effects of evolution and belongs not to them but to society. On this basis the idle man owes his idleness to society, the stupid man his stupidity, and the dishonest man his dishonesty. Nobody is responsible for anything, and all connection between conduct and the natural results of it is brought to an end.

Evolutionist emphasis on the environment as the controlling factor in the living world has long coloured political thought, and particularly socialist thought. Ability is assumed to be evenly distributed throughout the community and ready to spring forth in all directions with improved environment. When we turn to the actual facts as to the occurrence of ability, we find a different story. A handy summary of the results of various lines of investigation into the subject is Mr. Eldon Moore's *Heredity Mainly Human* (1934). Mr. Moore, on one side, reviews the facts revealed as to the distribution of ability by intelligence tests of school children. These tests are designed to discover the actual intelligence of the individual as measured by the time taken to solve various ingeniously devised problems. At the other end are various enquiries into the occurrence of exceptional ability, based on examination of the biographies of great men of history, etc.

The intelligence tests, according to Moore and Professor Terman, show that about three-quarters of the children tested are bunched together within 15 per cent. of the average, above and below; and that a third of the total tested come within 5 per cent. of average intelligence either way. About one in ten ranges from 15 to 25 per cent. above average intelligence, and no more than two or three individuals in a hundred have intelligence in excess of this. So far as they go, these tests thus suggest that the community is not supplied with vast concealed reserves of surplus ability. A further fact revealed by the tests is that, taken generally, the intelligence of the children corresponds closely to the occupational and social status of the parents. On top of this, tests of children reared under identical conditions in orphanages, etc., and many of whom have never known their parents, are stated by Mr. Moore to show just the same distribution of intelligence according to the occupational status of the parents as do home-reared children. This indicates that heredity and not environment is the dominating factor, and that what the scientists have noted about living organisms generally applies with equal force to human beings. It further indicates that whatever its other defects, the social and economic organisation under which we have lived has been efficient on the whole in allowing ability to find its level.

Turning now to the inquiries into the occurrence of exceptional ability, we find Mr. Moore's analysis of a number of different investigations showing that exceptional ability runs very largely indeed in strains. For example, one of the first of these investigators, Sir Francis Galton, in his *Hereditary Genius* (1869) made a list of 451 greatest men of all time in eight different fields of endeavour. He then discovered that these 451 men had sprung from a mere 300 families, which families had produced a further 562 men of eminence. Nearly a third of the 451 greatest men had eminent fathers, whereas only about one person in four thousand in the general population had a father reaching the degree of eminence adopted in the inquiry.

Another inquiry, made by Gunn, selected from the thirty thousand biographies of the British *Dictionary of National Biography* some 200 greatest men in the period from 1500 to 1900. Of these 200 greatest men, 21.5 per cent. had fathers with biographies also in the *Dictionary*, 30 per cent. had brothers, and 64 per cent had sons. Only 31.5 per cent. had no such kin with dictionary biographies. In some walks of life family influence of course, can do much in helping mediocrity along to prominence. Family influence, however, cannot make a man into a great or even eminent poet, artist, musician, philosopher, or scientist, and Mr. Moore notes that the great men in these spheres had just as many eminent relatives as had those in such careers as the law, the army, public life, etc., where influence could operate.

Both lines of investigation as to the occurrence of ability in human beings thus demonstrate heredity as the dominating factor. On top of this Mr. Moore in his book (p. 204) notes another influence co-operating to marked extent. He remarks that "artists, musicians, scientists, and poets beat their illustrious brethren hollow in the number of eminent sons; and they are the men whose gifts are most individual, who live and work at home, and make the closest contact with their children. A pedigree examination which cannot be reduced to figures, further confirms me in concluding that the one environmental circumstance of material importance in these highest grades of achievement is—Family Tradition."

Both lines of investigation are in complete harmony with the report of the scientists that a living organism is what it is by virtue of the germ plasm inherited from its parents. Added to this we have the mass of matter the diffusionist anthropologists have been assembling, all pointing to civilisation as founded on the discoveries and initiative of a very small number of exceptional men. The inference is that a social and economic organisation which gives free play to initiative and ability wherever found is the type of organisation which is likely to result in the highest civilisation. Evolutionist philosophy as applied to politics has tended to exactly opposite conclusions, and is the basis of present-day ideas of a completely controlled economy with everything regimented from above, and with little or no room for initiative from below.

As like tends to produce like, and the different kinds of degrees of ability run largely in strains, it follows that a social organisation under which the son slips easily and naturally into his father's place is one most likely to have the highest degree of efficiency and stability —provided that at the same time the organisation has sufficient flexibility to enable exceptional ability at all times to push through to its level. Furthermore, what Mr. Moore remarks about the influence of family tradition applies much more widely than to artists, musicians, and the like. It suggests strongly that the family business undertaking with son succeeding father is likely to be about the best type of undertaking with highest traditions and efficiency. Under present-day ideas this type of business appears to have been specially selected by those in authority for extinction. The existing financial system has tended for a long time to entrap such undertakings into debt and then hand over their mangled remains for incorporation in some financier-controlled combine. Planning and control ideas further accelerate the process.

Summed up, the indications are that the application of political programmes based on evolutionist philosophy is more likely to plunge our civilisation into the stagnation and decay in which every other civilisation has sunk than to lift it to higher levels. In this, as in every sphere it touches, the false theory of evolution is potently at work as an agent of disintegration.

Chapter XII

CONCLUSION

SIX years ago a small group of scientific men and others launched an Evolution Protest Movement in London. In a circular they issued it was pointed out that subversive doctrines were undermining every side of national life, and that this pointed to some fundamental fallacy operating on the national mind as a whole. This fallacy they believed to be the acceptance as true of the theory of evolution and its employment as the spring of action in all spheres. The reader is now in possession of the necessary material for forming his own opinion as to the soundness of this contention. He has viewed the small substratum of fact on which this top-heavy theory has been erected. He has also traced the peculiar origins of the theory : has seen how observed facts on which scientists are now dwelling run in flattest contradiction of it : and has glanced briefly over some of its principal fruits in the spheres of theology, morals, politics and economics.

The outstanding fact about the evolution theory is that it was a revival in the middle of the nineteenth century of ideas which had formed part of the intellectual ferment leading to the French Revolution. The propagation of evolution in the universities of Britain and elsewhere has been accompanied by an exactly similar growth of atheistical and revolutionary thought to that preceding the upheaval in France a century and a half ago. The weight of evidence throughout as we have seen, is that evolution is a theological and not a scientific product. If belief in God is rejected it becomes necessary to provide some theory as to the origin of life. The theory thus provided is next bolstered up with such material as can be scraped together, in great part by reckless distortion of observed fact, and then in turn becomes the means of further propagation of atheism. The result is that immense numbers of people are swung adrift from their bearings, social tradition is weakened, and upon one set of false assumptions an endless series of further false assumptions is erected in every direction. The entire theory and its offspring are products of the forces of darkness and not of light.

Modern evolutionist thought without doubt had its birthplace in the atheistical and revolutionary philosophy of eighteenth century France. Nothing is more mistaken than to regard the French Revolution as the spontaneous uprising of an oppressed people. All the evidence points to its having been a most carefully prepared event by men of great, though diabolical, intelligence in whose hands the mob were mere pawns. One of the greatest historians of the nineteenth century was Lord Acton, and he wrote of the French Revolution : "The appalling thing is not the tumult but the design. Through all the smoke and fire we perceive evidence of calculating organisation. The managers remain studiously concealed and masked, but there is no doubt about their presence from the first" (vide. *The Cause of World Unrest*).

Many writers, such as Mrs. Nesta Webster in her *World Revolution*, and the London *Morning Post* in its book of reprinted articles of 1920, *The Cause of World Unrest*, have gone into the nature of the hidden forces behind the French Revolution and other revolutions. The indications are that there has existed down through the centuries from ancient days a body of thought opposed to the whole World Order, and from time to time inspiring bloody upheaval. In its broadest aspects the subject was discussed by Professor Gilbert Murray of Oxford in his essay on *Satanism and the World Order* published in 1920. Professor Murray pointed out that both by thinkers and writers of pagan Greece and later on by those of Christendom the belief has been held that there exists a Cosmos or Divine Order : that what is good is in harmony with this Order, and what is bad is in discord against it. Opposed to this, there has also existed a belief that the World Order is evil and a lie. After noting that an appalling literature of hatred is in existence, dating at least from the eighth century B.C., Professor Murray added : "The spirit I have called Satanism, the spirit of unmixed hatred towards the existing World Order, the spirit which rejoices in any great disaster to the world's rulers, is perhaps more rife today than it has been for a thousand years. It is felt to some extent against all ordered Governments, but chiefly against all imperial Governments, and is directed more widely and intensely against Great Britain than against any other power."

This idea is exactly that which the late Lord Sydenham, a former Governor of various parts of the Empire, expressed in his autobiography *My Working Life* published in 1927 : "That the main bulwark of law and order and of Christianity should be laid low by any and every means is, therefore, the main object to which all revolutionary forces are now directed. The rest would be easy. The Union Jack is the most formidable enemy of the Red Flag."

A remarkable book first published in 1935 takes the view that the basic conflict in the world is between Supernaturalism, in which all power and authority is viewed as derived from God on high, and Naturalism, in which all power is viewed as derived from man below. The latter view leads to deification of man, which was the essence of the revolutionary philosophy of Rousseau : and under it no eternal principles of right and wrong exist, and murder—for example—ceases to be a crime if a victorious majority at an election so decrees. This book, *The Mystical Body of Christ in the Modern World* by the Rev. Dr. Denis Fahey, professor of philosophy and Church history at Blackrock College, Dublin, points out that having rejected Christ, the Supernatural Messias, the Jews thereafter looked forward to a Natural Messias and the establishment of a World Order under the Jewish nation. In this author's view all who do not fully accept the Supernatural Messias are inevitably drawn, consciously or unconsciously, into the army which is working for the advent of the Natural Messias. This line of thought is similar to the idea expressed by a Jewish author, Mr. Magnus Hermansson, who in his book, *Where Now, Little Jew*? contends that the Jewish question will only be solved by both Jews and Christians adopting Christianity, a view which was not shared by the American *Hebrew* of May 20, 1938, in recommending both Jew and non-Jew to read Mr. Hermansson's book. This again takes us back to Lord Bryce's diagnosis that the trouble with the world is that the nations have professed Christianity without practising it.

It is at least certain that the revolutionary upheavals have not made the world a better place to live in. Dr. Fahey in his book (p. xxi) quotes a spokesman at the Assembly of the French Grand Orient in 1920 as saying: "Every revolution had for object to bring about universal happiness. When our ancestors proclaimed the principle of Liberty, Equality, and Fraternity, they aimed at realising this state of happiness. After one hundred and thirty years we see the result of their efforts, and they are not famous. Of Liberty, there is not a shred left; of Equality, there is scarcely a trace; of Fraternity, there has never been a sign."

It has been remarked that any human society will always and inevitably form itself into the shape of a pyramid, and that if the pyramid is overturned the units in the human ant-hill will immediately and necessarily build up another pyramid without a moment's delay in order to preserve their social existence. All that the people ever get out of a revolution is a change of masters. They may exchange a Tsar for a Stalin and an aristocracy of nobles for an overlordship of Bolshevik commissars, but they will never escape from the pyramidal organisation of their social machinery.

This point was discussed by Nicholas Berdyaev in his essays on *The Russian Revolution* (1931). The author was an eye-witness of the Russian Revolution up to 1922 as a Professor in the University of Moscow, but then abjured Bolshevism and all its works and went into exile. He points out that atheistic communism is Christianity turned inside out, and either Christians must put their Christianity into practice or see the world reorganised in the name of a godless collectivity. Christianity, he says, is the only basis on which a solution can be found for the painful conflict between personality and society, which communism resolves in favour of society completely crushing out personality. "And," Berdyaev adds, "it is also the only basis on which a solution can be found for the no less painful conflict between the aristocratic and democratic principles in culture, resolved by communism in favour of completely overthrowing the aristocratic principle. On a basis of irreligion, either aristocracy oppresses and exploits democracy, or democracy vulgarises the souls of men, lowers the cultural level, and destroys nobility."

"Good which does not work itself into life, which has turned into conventional rhetoric so as to hide actual evil and injustice," says Berdyaev, "cannot avoid raising revolt, and righteous revolt, against its own self. The Christians of our bourgeois epoch of history have created most painful associations in the minds of the working class; they have not done Christ's mission to the souls of the oppressed and exploited . . . The situation of the Christian world face to face with communism is not merely that of the depositary of eternal and absolute truth, but also that of a guilty world which has not practised the truth it possesses, but rather turned traitor to it. Communists practise their truth and can always oppose that fact to Christians. Of course, Christian truth is much harder to carry out than communist truth. Much more, not less, is demanded of Christians than of communists, of materialists. And if Christians carry out less, and not more, Christian truth itself is not to blame." In Berdyaev's view, either the world must be renewed "in the name of God and Christ, of the spiritual principle in man, or in the name of divinised matter, in the name of a divinised human collectivity, in which the very image of man disappears and the human soul expires."

An enormous background lies behind the theory of organic evolution, our present subject. It has been noted that when Darwin published his *Origin of Species* he had apparently by no means fully convinced himself of its soundness. The fact of the matter is that after reflecting on the subject for over twenty years Darwin finally rushed into print in order to avoid being forestalled. Once his book had appeared and been acclaimed he cast all doubt aside and upheld his theory. Before publication all was uncertainty in his mind. For instance, in 1856 one finds him writing thus to his closest friend, Hooker, the botanist : "It is a melancholy, and I hope not quite true view of yours that facts will prove anything . . . I do not fear being tied down to error, i.e., I feel I should give up anything false published in the [proposed] preliminary essay, in my larger work; but I may thus, it is very true, do mischief by spreading error, which I have often heard you say is much easier spread than corrected" (Darwin's *Life and Letters*, ii, 70).

Recent events remind us that in the public schools of New Zealand evolution and Christianity are very differently regarded. In bygone years a number of leading atheists, some of whom came to occupy high positions, took advantage of the dissensions of the Christian sects cleverly to engineer a movement to de-Christianise the schools as a step to universal de-Christianisation.

In October, 1940, the newspapers recorded the Director of Education as officially notifying the Wellington Education Board that its decision to open the schools in its district by recital of the Lord's Prayer was entirely unlawful and of no effect. The Board, however, was reported as adhering to its decision. Turning to the Government *Syllabus of Instruction for Public Schools* issued in 1937—which repeated what had been there for many years—one discovers that whereas the doors of the State schools are kept tight shut against the eternal truths of Jesus Christ they are flung wide open to the manifold errors of Charles Darwin. The highly potent de-Christianising theory of organic evolution is laid down as one of the subjects to be taught to schoolchildren under the compulsory education system of New Zealand. Under the heading "Nature Study and Elementary Science." the syllabus says :

"The material for this subject is practically inexhaustible in that it comprises the whole of the animate population of the world together with the physical setting into which the many organisms have been born, and in which they have fought and are fighting their way upward to higher and yet higher stages of development. Ultimately Nature-Study should aim at enabling Man to understand and appreciate to some extent the scale of the universe and his own place in it" (pp. 42-3).

" . . . The scheme should provide for progressive treatment of the subject as the pupils advance in their school life, and in the higher classes the pupils should gain some definite ideas of the principle of evolution" (p. 43).

In passing, it may be noted that the New Zealand public school syllabus is not altogether up-to-date in its idea of evolution. It visualises living things fighting their way in the struggle for existence to "higher and yet higher stages of development." Darwin, it is true, concluded the *Origin of Species* with a picture of "progress towards perfection" by natural selection. Modern evolutionists are now satisfied to discover grounds for imagining evolution in any direction, upwards or downwards, sideways or forwards, purposeful or purposeless. In the Oxford University book on *Evolution* (p. 125) we find Professor A. M. Carr-Saunders writing as follows :

"The course of evolution has generally been downwards The majority of species have degenerated and become extinct, or what is perhaps worse, have gradually lost many of their functions. The ancestors of oysters and barnacles had heads. Snakes have lost their limbs and penguins their power of flight. Man may just as easily lose his intelligence."

This learned professor is director of the London School of Economics, founded by Mr. Sidney Webb and his socialist Fabian Society and expanded with money obtained through Lord Haldane from Sir Ernest Cassel, international financier, which endowment

Lord Haldane told Mr. J. H. Morgan, K.C., had been provided "to raise and train the bureaucracy of the future Socialist State (vide *Quarterly Review*, January 1929). In addition to supervising this undertaking, Professor Carr-Saunders in the essay quoted above reveals himself as possessed of plans for human evolution. He says the "less well-endowed" sections of the population are breeding too freely today and are far ahead of the "better endowed" sections. This evil can, in his opinion, be cured "once a cheap, efficient and simple contraceptive is available." This will enable the poor to behave the same as the "better endowed" do. Professor Carr-Saunders's idea is that when the lower orders are cured of the habit of having children, the upper-crust can be encouraged to have them on patriotic grounds, and upward evolution can then begin—that is, presuming that no other nation happens to take a fancy to possessing the depopulated country in the meantime. The big immediate step in human evolution, according to this professor, seems to be get the lower classes into the abortion parlours, etc., without delay.

While this book is in the press further comment on the communistic leanings of New Zealand intellectuals comes to hand in the December, 1940, issue of a little Auckland publication, *View*. Some leading lights of Auckland University College had protested against the efforts of the Auckland Education Board to keep the primary schools clear of communistic teachers, and *View* said : "Whatever may be the value . . . of the operations of the University in the field of natural science, the prevailing trend of its influence in the sphere of human affairs—the 'social sciences'—is somewhat worse than worthless. Its deliverances fall below the commonsense of the average man. This does not apply to New Zealand alone, but to most of its sources of inspiration and recruitment overseas."

If *View* will dig deeper still it will find that the essential worthlessness of present-day university teaching is that natural science has been made a vehicle for atheistic and materialist propaganda per medium of the imbecilities of the evolution theory. The modern university college is a machine for de-Christianising and demoralising the community.

APPENDIX

SCIENTISTS WHO REJECT EVOLUTION

Evolutionists commonly make statements leading the casual reader to believe that all scientists accept evolution as established fact. If these statements are attentively read, however, almost all of them will be found to contain an unobtrusive qualification. For instance, in the current fourteenth edition of the *Encyclopaedia Britannica* the article on evolution says that "among competent biologists and geologists there is not a single one who is not convinced," etc. This means no more than that in the opinion of the writer of the article the scientists who reject evolution are not "competent." Similarly, when the council of the American Association for the Advancement of Science in 1922 proclaimed by resolution that "the evidences in favour of the evolution of man are sufficient to convince every scientist of note in the world," they are likewise merely throwing dust in the public's eyes. All that the announcement means is that this scientific body is blacklisting the scientists who reject evolution and is refusing to regard them as of "note." This strain runs through evolution from top to bottom.

The following list of scientists who have definitely rejected the entire theory of organic evolution is compiled from a pamphlet by Lieut. Col. L. Merson Davies, from a leaflet issued by the Evolution Protest Movement, and from names mentioned in Mr. Douglas Dewar's books :

PHYSICISTS

Sir J. Ambrose Fleming, F.R.S.

President of the Victoria Institute and Philosophical Society of Great Britain, has been awarded many medals and honours by various scientific societies; inventor of the thermionic valve making radio broadcasting possible; has flatly rejected the entire theory of evolution in numerous addresses.

Louis Trenchard More

Professor of physics, University of Cincinnati, U.S.A., an expert physicist who has ridiculed evolution in his *Dogma of Evolution* (1925).

ZOOLOGISTS

Albert Fleischmann, GR.

Professor of Zoology and Comparative Anatomy in the University of Erlangen, Germany, a scientist of European reputation: roundly attacked evolution in 1901 in his book, *Die Descendenztheorie* (1901), and remained completely unmoved by the abuse heaped upon him; in a letter to Col. Merson Davies in 1931 said : "I reject evolution because I deem it obsolete; because the knowledge, hard-won since 1830, of Anatomy, Histology, Cytology and Embryology, cannot be made to accord with its basic idea"; attributes persistence of evolution to "mankind's love of fairy tales."

L. VIALLETON

Professor of Zoology, Anatomy, and Comparative Physiology at Montpellier University, France, member of the Royal Academy of Science of Turin (which marks him as a leading European scientist); attacked evolution in his *Morphologie Generale* (1924); his book *L'Origine des Etres Vivants* appeared in 1929 and ran through fifteen editions by 1930, but being strongly against evolution no English translation ever appeared.

E. G. DEHAUT

French biologist and palaeontologist, author of numerous scientific works, professes his belief in intervention of creative power to produce new types.

DOUGLAS DEWAR, F.Z.S.

An authority on Indian birds, rejected evolution in 1931; and has since written the following books condemning it: *Difficulties of the Evolution Theory, More Difficulties of the Evolution Theory, Man, a Special Creation* and *A Challenge to Evolutionists:* the latter being a report of his share of a debate (as representative of the Evolution Protest Movement) with Mr. J. J. McCabe (representing the Rationalist Press Association), who threatened legal proceedings if his part of the debate were published.

GEORGE BARRY O'TOOLE

A Catholic lecturer or professor of biology; author of *The Case against Evolution*, published by the Macmillan Co., New York, in 1931.

VINCENZO DIAMARE

Director of the Institute of Osteology and General Physiology in the University of Naples, rejected evolution in a book published in 1912.

D. CARAZZI

Another Italian biologist quoted by Vialleton, rejected evolution in his *Il Dogma dell' Evoluzione*, 1920.

GIULIO FANO

Director of the Institute of Osteology and General Physiology in the University of Rome; attacked evolution in his book, *Brain and Heart*, of which an English translation was published by the Oxford University Press in 1926.

BOTANIST

JOHANNES REINKE, GRR.

Emeritus Professor of Botany at Kiel University, Germany; has published many papers attacking evolution; he and Professor Fleischmann, hold rank in Germany equivalent to about that of Privy Councillors in England.

GEOLOGISTS

PAUL LEMOINE

Past president of the Geological Society of France and director of the Museum d'Histoire; describes evolution as "a sort of dogma in which its priests do not believe, but which they uphold before the people" (*vide* Dewar's *More Difficulties of the Evolution Theory*).

W. BELL DAWSON, D.SC., F.R.S.C.

A well-known Canadian geologist and a Laureate of the French Academy of Sciences; like his father, Sir J. W. Dawson, F.R.S., former principal of McGill University, Canada, he is a determined opponent of evolution; author of *The Bible Confirmed by Science*, and various pamphlets including *Is Evolution True?* No 1 to 5.

G MCCREADY PRICE

Professor of Geology, and author of *The Phantom of Organic Evolution*.

LT.-COL. I. MERSON DAVIES, F.R.S.E., F.R.A.I., F.G.S.

A palaeontological research worker specialising in foraminifera, who states that he is constantly face to face with facts regarding the fossil faunas of the past which he is unable to reconcile with the theory of evolution.

ARCHAEOLOGIST

SIR CHARLES MARSTON, F.S.A.

Vice-chairman of the British School of Archaeology in Egypt, member of the executive of the Palestine Exploration Fund, and collaborator with Professor Garstang in the excavation of Jericho; author of *The Bible is True*.

The above list is not exhaustive, but it is sufficient to show that when evolutionists state that no competent biologists or geologists, or no scientists of note, disbelieve in evolution they are not telling the truth. The information at hand does not disclose the religious beliefs of all the above-listed scientists. Professor Fleischmann is stated to be an agnostic, and Professor L. T. More appears from his remarks to have no respect for the authority of the Bible. Sir Ambrose Fleming, Sir Charles Marston, Professors Reinke and McCready Price, Dr. Bell Dawson, Lieut.-Col. L. Merson Davies, Mr. Douglas Dewar, and the Rev. G. Barry O'Toole are listed as of Christian belief. There are many other scientists who do not believe in evolution but have not so far publicly rejected it.

NOTES

NOTES

NOTES

NOTES

NOTES

NOTES

If you have enjoyed this book, consider making your next selection from among the following . . .

Prices guaranteed through June 30, 1994.

At your bookdealer or direct from the publisher.

Prices guaranteed through June 30, 1994.